Collins

11+
Maths

Practice Papers
Book 2

Flora MacInnes

Introduction

The 11+ tests

In most cases, the 11+ selection tests are set by GL Assessment (NFER), CEM or the individual school. You should be able to find out which tests your child will be taking on the website of the school they are applying to or from the local authority.

These single subject practice test papers are designed to reflect the style of GL Assessment tests, but provide useful practice and preparation for all 11+ tests and common entrance exams.

The score achieved on these test papers is no guarantee that your child will achieve a score of the same standard on the formal tests. Other factors, such as the standard of responses from all pupils who took the test, will determine their success in the formal examination.

Collins also publishes practice test papers, in partnership with The 11 Plus Tutoring Academy, to support preparation for the CEM tests.

Contents

This book contains:

- four practice papers – Tests A, B, C and D

- a multiple-choice answer sheet for each test

- a complete set of answers, including explanations.

Further multiple-choice answer sheets can be downloaded from our website so that you can reuse these papers: collins.co.uk/11plus

Maths

Mathematics tests are used by schools to assess the ability of each child and determine whether they have attained the required standard of mathematical skills, reasoning and problem-solving.

It is particularly important to provide maths practice as the 11+ tests may test skills that are slightly more advanced than those on the national curriculum for your child's age.

The importance of practice

Practice will help your child to do his or her best on the day of the tests. Working through a number of practice tests allows your child to practise answering a range of test-style questions. It also provides an opportunity to learn how to manage time effectively, so that time is not wasted during the test and any 'extra' time is used constructively for checking.

Getting ready for the tests

If your child is unfamiliar with mathematics 11+ papers, it may be advisable to attempt a few questions first without time constraints and give your child the opportunity to ask questions and receive some initial feedback.

It is best to do the tests at a time when your child is alert and able to concentrate fully on them. Tiredness and other distractions will have an adverse effect on their performance. Spend some time talking with your child before the test so that they understand the purpose of the practice papers.

It is also good to go through with your child some tactics to adopt when attempting the paper. These might include:

- Work quickly and carefully through the questions.

- All the questions are worth equal marks, so don't spend too long on any one question.

- If you get stuck, leave it and then come back to it if you have time.

- If you have spare time at the end, go back and check your answers. Every mark counts!

Administering the tests

Make sure that the surroundings are appropriate and quiet. Your child will need a pencil and rubber and some paper for rough working. A calculator must not be used.

Allow your child some time at the start to read the information on the front of the paper.

Each mathematics test consists of 50 questions to be completed in 50 minutes. It is essential that your child is able to work uninterrupted for this time. A clock should be provided so that a check can be kept on the time left.

Multiple-choice tests

For this style of test, the answers are recorded on a separate answer sheet and not in the book. This answer sheet will often be marked by a computer in the actual exam, so it is important that it is used correctly. Answers should be indicated by drawing a clear pencil line through the appropriate box and there should be no other marks. If your child indicates one answer and then wants to change their response, the first mark must be fully rubbed out. Practising with an answer sheet now will reduce the chance of your child getting anxious or confused during the actual test.

Marking

Award one mark for each correct answer. Do not award any marks for correct working with an incorrect answer, or any half-marks.

It is important that you start by providing some positive feedback for questions that have been correctly answered. This will help your child to identify the topics that they are confident with. Next, identify questions where your child has made an easily correctable mistake or misread the question. Ask your child to try these questions again to see if correct answers can be obtained. Finally, identify the questions that your child provided incorrect answers for, or was unable to answer at all. Revise the material covered by these questions and re-attempt them.

And finally…

Let your child know that tests are just one part of school life and that doing their best is what matters. Plan a fun incentive for after the 11+ tests, such as a day out.

Contents

ACKNOWLEDGEMENTS

The author and publisher are grateful to the copyright holders for permission to use quoted materials and images.

Every effort has been made to trace copyright holders and obtain their permission for the use of copyright material. The author and publisher will gladly receive information enabling them to rectify any error or omission in subsequent editions. All facts are correct at time of going to press.

Published by Collins
An imprint of HarperCollins*Publishers* Limited
1 London Bridge Street
London SE1 9GF

HarperCollins*Publishers*
Macken House, 39/40 Mayor Street Upper,
Dublin 1, D01 C9W8, Ireland

ISBN 9780008278021

First published 2018
This edition published 2020
Previously published as Letts

10 9 8 7

©HarperCollins*Publishers* Limited 2020

British Library Cataloguing in Publication Data.

A CIP record of this book is available from the British Library.

Commissioning Editors: Michelle I'Anson and Alison James
Author: Flora MacInnes
Project Management: Fiona Watson
Cover Design: Sarah Duxbury and Kevin Robbins
Production: Natalia Rebow
Printed and bound in the UK

 MIX
Paper | Supporting responsible forestry
FSC™ C007454

This book is produced from independently certified FSC™ paper to ensure responsible forest management.

For more information visit: www.harpercollins.co.uk/green

Mathematics
Multiple-Choice
Practice Test A

Read the following carefully.

1. You must not open or turn over this booklet until you are told to do so.

2. This is a multiple-choice test, which contains a number of different types of questions.

3. You should do any rough working on a separate sheet of paper.

4. Answers should be marked in pencil on the answer sheet provided, not on the test booklet.

5. If you make a mistake, rub it out as completely as you can and put in your new answer.

6. Work as carefully and as quickly as you can. If you cannot do a question, do not waste time on it but go on to the next.

7. If you are not sure of an answer, choose the one you think is best.

8. You will have 50 minutes to complete the test.

1. Mr Kumar can type 47 words per minute. If he averages this exact speed, how many words can he type in 2 hours?

   A 94   B 9400   C 6674   D 5880   E 5640

2. Farmer Bunce puts 1492 eggs into boxes. Each box holds 6 eggs. How many boxes does he need?

   A 237   B 249   C 252   D 248   E 244

3. Dev thinks of a number, triples it and adds 14. The answer is 125. What was the number Dev first thought of?

   A 39   B 40   C 43   D 36   E 37

4. 300 workers were asked which food they preferred for Friday work treats. This pie chart shows the results.

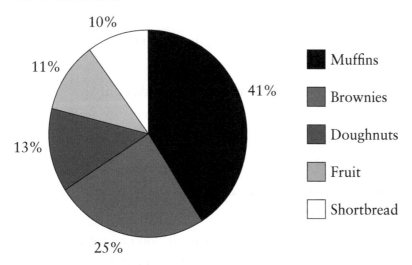

   How many workers preferred doughnuts?

   A 130   B 13   C 25   D 37   E 39

5. Two strikers play for Belting United. Last season they scored a total of 125 goals. Striker A scored 17 more goals than Striker B. How many goals did Striker B score?

   A 51   B 57   C 54   D 52   E 55

6. Barry has square tiles that are 20 cm wide. He wants to cover his bathroom wall. If the bathroom wall is 2 m by 3 m, how many tiles does he require to cover the whole wall?

   A 20   B 60   C 160   D 150   E 80

7. The bar chart shows how many hours of sport Wesley played in September.

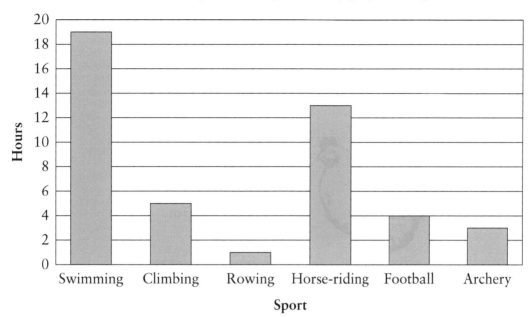

What was the average time per day that Wesley spent playing sport in September?

A 0.5 hours   **B** 3 hours   **C** 2.25 hours   **D** 1.5 hours   **E** 2.5 hours

8. What is the value of the 1 in this number?

5120528

A 1000   **B** 100000   **C** 10000   **D** 15000   **E** 1000000

9. A company has 420 employees. 85% of employees drive to work. $\frac{1}{12}$ of employees cycle to work. The remainder walk to work. How many employees walk to work?

A 36   **B** 24   **C** 28   **D** 48   **E** 32

10. How many hundredths must be added to 2.13 to make 3?

A 970   **B** 87   **C** 0.97   **D** 0.87   **E** 870

11. Mr Fox builds a chicken coop. The coop is 13 m by 15 m. Mr Fox wants to fence the outside of the coop. A metre of wire fencing costs £2.67. How much will it cost him to fence the coop?

A £40.05   B £520.65   C £149.52   D £187.69   E £231.58

15 m

13 m

12. The two angles y and $8y$ form a straight line.

y $8y$

What is the value of y?

A 15°   B 10°   C 20°   D 180°   E 120°

13. Look at this pictogram.

Number of trains		
Platform	Key: 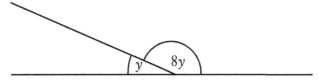 represents 16 trains	
A	🚂🚂🚂🚂🚂🚂🚂	
B	🚂🚂	
C	🚂🚂🚂🚂	

How many more trains leave from Platform A than Platform C?

A 56   B 44   C 58   D 48   E 62

14. A tortoise travels 0.4 cm in 20 seconds. What is the tortoise's average speed in metres per hour?

   A 1.2   B 7.2   C 72   D 0.72   E 12

15. Gabriel makes 17 litres of apple juice and pours it into 125 ml bottles. How many bottles can he fill?

   A 102   B 215   C 137   D 136   E 224

16. What would you multiply 81430 by to get the answer 81.43?

   A 1000   B 0.01   C 0.0001   D 100   E 0.001

17. Sarah and Barry go to Potato Café. The cost of a jacket potato is £1.50 and the cost of toppings is 30p per topping. Calculate the total cost of their meal if Sarah has 1 potato and 4 toppings and Barry has 2 potatoes and 3 toppings.

   A £4.50   B £6.60   C £5.80   D £5.85   E £6.70

18. Monica buys 2.548 kg of dog food for her dog, Roxy. Roxy eats 26 g of food twice per day. For how many weeks will the dog food last?

   A 4 weeks   B 7 weeks   C 5 weeks   D 9 weeks   E 10 weeks

19. Calculate $340 \div (85 \times 0.2)$

   A 2   B 4   C 20   D 8   E 12

20. Samira decides to pave a path around the edge of her rectangular garden. The lawn in the middle measures 225 cm by 7 m. The width of the path is a uniform 1850 mm all the way round.

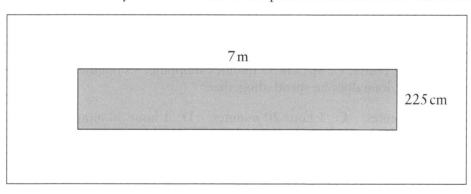

7 m

225 cm

   What is the outer perimeter of Samira's path?

   A 33.03 m   B 32.15 m   C 28.65 m   D 29.5 m   E 33.3 m

21. Mr O'Murphy buys x number of tickets. Mr Bailey buys twice as many tickets as Mr O'Murphy. What expression would give the total cost of Mr Bailey's and Mr O'Murphy's tickets?

   A $x + 2y$ B $x + 2$ C $x - 2$ D $3x$ E $3x - 1$

22. Muffins cost 85p. If Tilly has £6, what is the greatest number of muffins she can buy?

   A 6   B 5   C 7   D 9   E 8

23. What is the 7th term in the sequence?

37, 29, 20, 10, −1, ..., ?

A −13　　　　　B −18　　　　　C 24　　　　　D 26　　　　　E −26

24. 387 children go on a school trip. Each bus holds 27 children and costs £20 to hire. What will the total cost be to hire enough buses for all the children?

A £240　　　　　B £320　　　　　C £300　　　　　D £280　　　　　E £340

25. Which of these pairs of numbers are equally distant from 5?

A 5.9 and 5.01　　B 5.11 and 4.99　　C 5.03 and 4.7　　D 5.91 and 4.91　　E 4.92 and 5.08

26. Look at the results of the men's Olympic long jump.

Country	Distance
USA	8.431 m
Brazil	8.41 m
Sweden	8.429 m
Russia	8.08 m
Spain	8.437 m

Which country came in 2nd place?

A USA　　　　　B Brazil　　　　　C Sweden　　　　　D Russia　　　　　E Spain

27. Which of these is closest in value to 4?

A 3.996　　　　　B 4.01　　　　　C 3.99　　　　　D 4.11　　　　　E 4.007

28. Sanjay goes to a shopping centre. He spends 35 minutes shopping, 45 minutes eating and 10 minutes parking. How long does he spend altogether?

A 1.3 hours　　B 80 minutes　　C 1 hour 20 minutes　　D 1 hour 30 minutes　　E 2 hours

29. The table below shows what is in 45 g of Yumwow chocolate.

Nutritional information per 45 g of Yumwow chocolate	
Protein	8 g
Carbohydrate	23 g
Fibre	2 g
Fat	7 g
Other	5 g

Nora draws a pie chart to show this information. She calculates the angles to the exact degree. What is the correct angle for Carbohydrate?

A 230°　　　　　B 1035°　　　　　C 138°　　　　　D 260°　　　　　E 184°

30. The time is thirteen minutes to ten in the evening. How would this be written in a 24–hour clock format?

    A 10:13   B 22:13   C 09:47   D 21:47   E 21:46

31. What percentage of £10 is 1p?

    A 10%   B 1.1%   C 1%   D 0.01%   E 0.1%

32. It is –16°C in Lapland. Tomorrow it will be 5° warmer. If the day after tomorrow is 12° colder than tomorrow, what will the temperature be the day after tomorrow?

    A –28°C   B –11°C   C –23°C   D –17°C   E 3°C

33. Xan measured out some flour. The line on the scale below shows how much he has measured.

    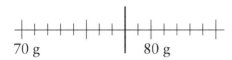

    70 g   80 g

    He accidentally spills 23 g of the flour he has measured. How much flour does he have left?

    A 101 g   B 51 g   C 95 g   D 55 g   E 56 g

34. The population of Latvia is one million, nine hundred and seventy thousand, five hundred and thirty. The population of Luxembourg is five hundred and seventy-five thousand, seven hundred and forty-seven. What is the difference in population between Luxembourg and Latvia?

    A 531983   B 1394783   C 378217   D 876513   E 981372

35. A two-pence coin weighs $7\frac{1}{4}$ grams. Ajeet has 68p in his wallet in two-pence coins. If his wallet weighs 230 g, what is the total weight of the wallet and the coins he has?

    A 723 g   B 1216 g   C 136 g   D 466 g   E 476.5 g

36. Kirsty and Stella are sharing a packet of sweets. There are 45 sweets in total. Stella gets twice as many sweets as Kirsty. Stella eats 13 and gives 17 away to her friends. How many does she have left?

    A 15   B 14   C 0   D 23   E 19

37. A train leaves Habington at 18:42 and arrives in Wellington at 21:34. How many minutes does the journey take?

    A 152 minutes   B 146 minutes   C 172 minutes   D 164 minutes   E 169 minutes

38. Which operation has changed 5601 to 5601000?

    A × 100   B × 10   C × 1000   D ÷ 100   E ÷ 10000

39. A pirate places his treasure at the point (3, 4). Which letter indicates the location of his treasure?

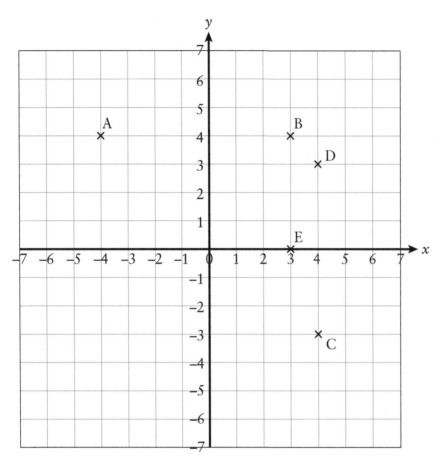

40. Mr Burton saves 50p every day. He begins saving on 1st March. In which month will he have saved £100?

A June   B July   C August   D September   E October

41. The parish council of Hallford has 73 members. The table shows some information about them.

	Male	Female	Total
Under 50			32
Over 50		21	41
Total		39	73

How many members of the council are male and under 50?

A 15   B 18   C 20   D 14   E 12

42. If I buy 4 packets of crisps at 47p each and 12 cartons of milk at 32p each, how much change should I get if I pay with a £20 note?

    A £13.78   **B** £14.28   **C** £6.86   **D** £10.74   **E** £12.36

43. Here are 5 lines.

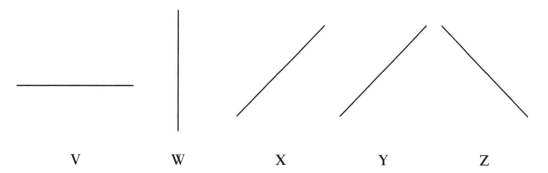

      V   W   X   Y   Z

Which one of the following statements is true?

    A Line Y is 180° in relation to line V.

    B Line W is perpendicular to line Y.

    C Line X is perpendicular to Line W.

    D Line Y is parallel to line Z.

    E Line V is perpendicular to line W.

44. Which of these shapes is not a hexagon?

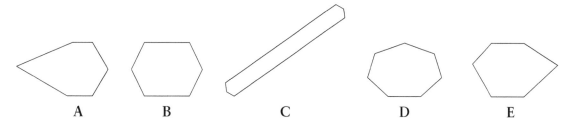

      A   B   C   D   E

45. $5c = b - 4$

Which of the following is not true?

    A $5c - 4 = b$

    B $5c + 4 = b$

    C $b - 5c = 4$

    D $5c - b = -4$

    E $10c = 2b - 8$

NOW GO ON TO THE NEXT PAGE

46. Zera has a regular octagon.

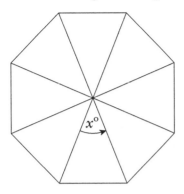

What is the size of angle x?

A 45° **B** 145° **C** 135° **D** 105° **E** 67.5°

47. Which one of these expressions is equivalent to $4(p + 6)$?

A $4p - 20$ **B** $4(p + 24) - 6$ **C** $4p + 6$ **D** $4p + 24$ **E** $4p + 10$

48. A rectangle is made by joining the points J, K, D and C together.

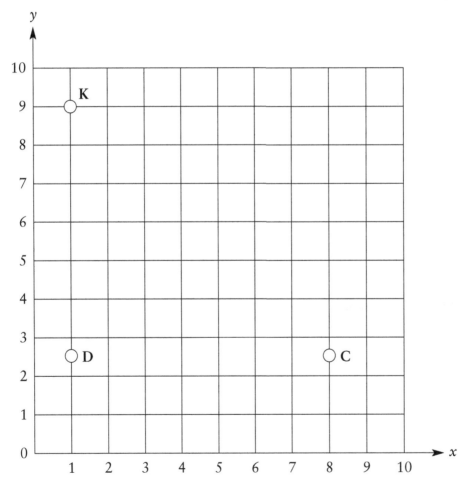

What is point J?

A (9, 8) **B** (8, 9) **C** (9, 9) **D** (9.5, 7) **E** (4, 10)

49. $61 \times 185 = 11285$

What is $112.85 \div 0.0185$?

A 6.1   B 610   C 0.61   D 6100   E 6.16

50. Wendy has a circular pond in her doll's house. Its diameter is 1.28 mm.

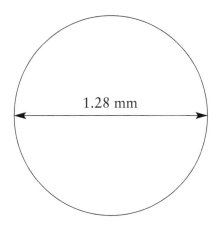

1.28 mm

If the pond's diameter was 35 times bigger than its current diameter, what would it measure in cm?

A 448 cm   B 4096 cm   C 5.64 cm   D 44.8 cm   E 4.48 cm

TEST ADVICE

This information will not appear in the actual test.
It is included here to remind you not to stop working
until you are told the test is over.

CHECK YOUR ANSWERS AGAIN IF THERE IS TIME

CORRECTING EVEN ONE MISTAKE CAN MEAN AN EXTRA MARK

Mathematics
Multiple-Choice
Practice Test B

Read the following carefully.

1. You must not open or turn over this booklet until you are told to do so.

2. This is a multiple-choice test, which contains a number of different types of questions.

3. You should do any rough working on a separate sheet of paper.

4. Answers should be marked in pencil on the answer sheet provided, not on the test booklet.

5. If you make a mistake, rub it out as completely as you can and put in your new answer.

6. Work as carefully and as quickly as you can. If you cannot do a question, do not waste time on it but go on to the next.

7. If you are not sure of an answer, choose the one you think is best.

8. You will have 50 minutes to complete the test.

1. Saima buys 6 candles priced at 38p each and 1 box of matches priced at 57p each. How much change should she receive from a £5 note?

   A £2.43   B £3.14   C £2.55   D £2.15   E £2.27

2. Liam goes to the cinema to watch *Star Wars*. The screening begins at 7.25 p.m. and the film lasts 116 minutes. There are 17 minutes of commercials before the film starts. What time does the film end?

   A 20:58   B 21:03   C 21:38   D 20:47   E 22:06

3. At the village fete, Mr Hopkins wins a prize for the heaviest carrot. The carrot weighs 2003 grams. What is the carrot's weight in kilograms?

   A 0.2003 kg   B 2.003 kg   C 2.03 kg   D 2.033 kg   E 2.3 kg

4. Which has the greatest value?

   A $\frac{1}{4}$ of 112   B $8\frac{1}{3}$ C $\frac{5}{8}$ of 48   D 31% of 105   E 20% of 145

5. The time on my clock is eight o'clock. What is the reflex angle between the clock hands?

   A 150°   B 230°   C 120°   D 270°   E 240°

6. Calculate 43 × 274 + 57 × 274

   A 31578   B 28980   C 27400   D 32434   E 284500

7. A chocolate gift hamper weighs 1.43 kg. The empty hamper weighs 450 g. The hamper contains 28 bars of chocolate. What is the weight of 1 bar of chocolate?

   A 20 g   B 27 g   C 40 g   D 35 g   E 21 g

8. What is the difference in temperature between the hottest city and the coldest city?

City	Temperature
Berlin	18°C
Havana	33°C
Mumbai	35°C
London	15°C
Moscow	–17°C
Reykjavik	–1°C

   A 43°C   B 52°C   C 21°C   D 34°C   E –6°C

9. Mr Ramirez thinks of a number, multiplies it by 3 and then subtracts 50. He divides the result by 4. The answer is 16. What was the number Mr Ramirez first thought of?

   A 24   B 30   C 28   D 38   E 44

10. Ade collects 284 seashells. He gives $\frac{1}{4}$ away to his brother, and then gives away $\frac{4}{6}$ of the remaining amount as presents. He accidentally loses 6 but sells the rest for 4p per shell. How much money does he make from selling the shells?

   A £13.02   B £8.88   C £1.24   D £3.44   E £2.60

11. Mrs Brown records the number of sandwiches she sells in her shop in a week.

Monday	48
Tuesday	49
Wednesday	41
Thursday	24
Friday	37
Saturday	50
Sunday	66

   What is the mean number of sandwiches she sells?

   A 49   B 45   C 48   D 37   E 41

12. Barnford Café sells hot chocolate made with hot chocolate mixture and milk for £2.95 per cup. The hot chocolate mixture costs 37p per cup. Each cup uses $\frac{1}{4}$ of a litre of milk. Milk costs 88p per litre. What percentage of the total selling price does the hot chocolate cost to make?

   A 25%   B 33%   C 21%   D 20%   E 38%

13. There are 2 patches of grass with exactly the same area. The length of patch A is 12 m and the width is 3 m. If patch B is 8 m in length, what is its width in cm?

   A 475 cm   B 4.6 cm   C 450 cm   D 510 cm   E 600 cm

14. Calculate 3 − 0.0038

   A 2.9962   B 2.62   C 2.0062   D 2.612   E 2.64

15. Below are the dates of birth of 4 children.

Janet	29th April 1992
Daniel	15th July 1992
Obie	8th December 1991
Ronald	30th November 1991

How many days older is the eldest child than the youngest child? 1992 was a leap year.

A 111   B 191   C 207   D 314   E 228

16. Faryaal thinks of a shape that is a quadrilateral. All of its sides are equal and it has equal opposite angles. Which shape is she thinking of?

A Trapezium   B Irregular hexagon   C Rhombus   D Rectangle
E Scalene triangle

17. Mrs Jules parks the car from 9.45 a.m. until 6.05 p.m. Parking costs 42p per hour, but is charged exactly to the minute. How much will Mrs Jules need to pay when she goes back to her car?

A £3.05   B £3.72   C £3.88   D £3.96   E £3.50

18. Tiles cost £1.36 per square metre. Tatiana wants to cover the whole wall with tiles. What is the total cost of tiles for a bathroom wall that is 3.5 metres long and 4 metres high?

A £18.84   B £15.76   C £16.66   D £19.04   E £19.44

19. What is the angle marked x in this triangle?

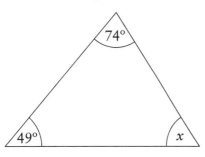

A 67°   B 76°   C 57°   D 45°   E 82°

20. What is 378.036 rounded to the nearest hundredth?

A 400   B 378.4   C 378.1   D 378.04   E 378.14

21. In Year 7, the ratio of boys to girls is 17:4. If there are 52 girls, what is the total number of pupils in Year 7?

A 262   B 195   C 273   D 259   E 287

22. An architect draws a plan of a kitchen. 1 cm in the drawing is equivalent to 7 m. If the room is 2.5 cm wide and 3.5 cm long on the plan, what is the perimeter of the kitchen?

    A 84 m   **B** 46 m   **C** 7200 cm   **D** 55 m   **E** 1005 cm

23. There are boxes that hold 6 candles. How many of these boxes are required for 105 candles?

    A 17   **B** 15   **C** 19   **D** 18   **E** 14

24. A map of Edder Park is drawn to a scale of 1:3420. What real distance in km is represented by 1 cm on the map?

    A 342 km   **B** 34.2 km   **C** 34200 km   **D** 0.342 km   **E** 0.0342 km

25. How much bigger is $7\frac{2}{5}$ than $4\frac{9}{10}$?

    A $3\frac{7}{10}$ **B** $3\frac{1}{10}$ **C** $2\frac{1}{2}$ **D** $2\frac{4}{5}$ **E** $2\frac{9}{10}$

26. If keyrings cost £2.68 each, how many can Jamil buy with a £20 note?

    A 7   **B** 8   **C** 9   **D** 6   **E** 11

27. Which answer has the greatest value?

    A $\frac{4}{5}$ of 950   **B** $\frac{9}{10}$ of 950   **C** 857   **D** 0.75 of 950   **E** $\frac{1}{2}$ of 950

28. If a car travels 56 kilometres per hour, how far will it travel in 3 hours 15 minutes?

    A 1820 m   **B** 18200 m   **C** 16200 m   **D** 18300 m   **E** 182000 m

29. Salim walks 3.05 km every day. His father walks 347 m less than him. How far does Salim's father walk every day?

    A 2703 m   **B** 2.0703 km   **C** 2658 m   **D** 3.397 km   **E** 3153 m

30. What is the 6ᵗʰ term in the sequence?

107, 98, 89, 80, ... , ?

    A 70   **B** 71   **C** 69   **D** 62   **E** 64

31. Mr Rodriguez drives 17 km in 15 minutes. If he continues at this exact speed, how far does he travel in 1.5 hours?

    A 170 km   **B** 102 km   **C** 106 km   **D** 138 km   **E** 107 km

32. How many fifths are there in 9?

A 9   B 5   C 95   D 99   E 45

33. 240 schoolchildren were asked what their favourite genre of literature was. This pie chart shows the results.

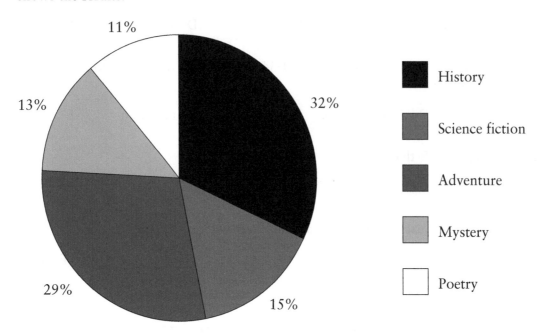

How many schoolchildren preferred science fiction books?

A 15   B 32   C 34   D 36   E 48

34. How many thousandths are there in 34.793?

A 34   B 3   C 743   D 3000   E 34793

35. 45 rolls cost £3.15 and 25 slices of ham cost £3.50. What is the cost of 2 rolls and 3 slices of ham?

A 69p   B 56p   C 44p   D 62p   E 58p

36. Look at this graph.

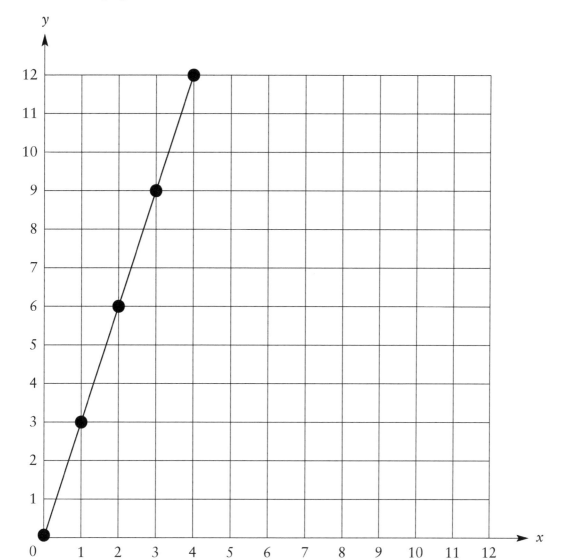

What is the equation of the line?

A $y = 3 + x$ **B** $y = 2x$ **C** $y = x - 3$ **D** $y = 3(x - 1)$ **E** $y = 3x$

37. Calculate $236 \div 16$

A $14\frac{1}{2}$ **B** $15\frac{2}{3}$ **C** $15\frac{3}{4}$ **D** $14\frac{3}{4}$ **E** $15\frac{1}{4}$

38. The bar chart shows the number of goals scored by Hammerville United each month.

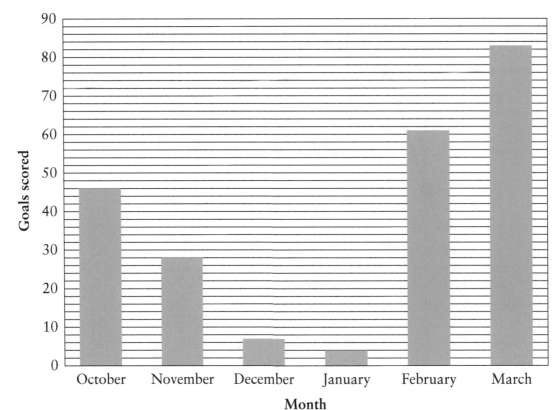

How many more goals did they score in February than in January?

A 34   B 22   C 45   D 57   E 29

39. If 16ᵗʰ March falls on a Friday, on which day of the week will 11ᵗʰ April fall?

A Wednesday   B Thursday   C Friday   D Monday   E Tuesday

40. On a map, 50 cm represents 20 km. What is the scale of the map?

A 5:2000   B 1:4000   C 1:400   D 1:40000   E 5:20000000

41. The graph below shows the conversion rate between yen and euros.

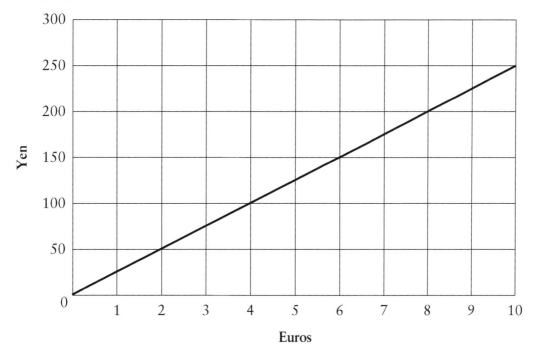

Using the graph, estimate how many yen I would receive if I changed 25 euros.

A 500   B 625   C 238   D 548   E 782

42. Farmer Boggins plants beetroot in his garden patch, which is shown in the plan below.

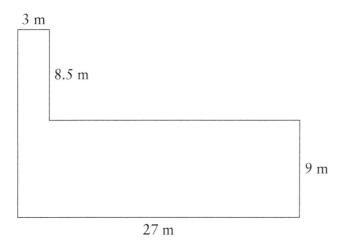

Work out the total area of his garden patch.

A 52.5 m²   B 286.5 m²   C 275.5 m²   D 268.5 m²   E 147.5 m²

43. What is 43.097 kg to the nearest 10 g, shown in kg?

A 43.09 kg   B 43.98 kg   C 43.098 kg   D 43.1 kg   E 43.01 kg

44. Daisy makes a huge pizza and cuts it into 6 equal slices. The whole pizza weighs 1.668 kg.

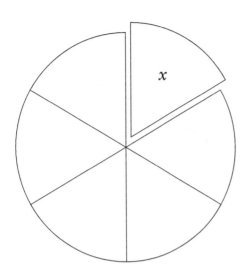

If Jeremy gets a slice (labelled *x*), and adds 12 individual tomato pieces, each weighing 13 g, what is the total weight of his pizza slice?

A 426 g   B 444 g   C 198 g   D 434 g   E 301 g

45. Calculate $\frac{4}{6}$ of $\frac{1}{3}$ of 432

A 324   B 284   C 96   D 86   E 285

46. 25 ice-creams cost £3.25. If I buy 31 ice-creams, how much change should I receive from £10?

A £4.03   B £7.53   C £5.07   D £5.97   E £6.31

47. Calculate 23% of 700

A 145   B 148   C 153   D 157   E 161

48. The mean of 5 numbers is 8. We know that 3 of the 5 numbers are 13, 8 and 13. The 2 other numbers are the same. What is the value of each of these 2 numbers?

A 7   B 13   C 8   D 3   E 5

49. The average temperature of Chillyville was recorded for 5 days during 1 week in December.

Here are the temperatures:

| 2°C | –4°C | –14°C | –7°C | 12°C | –15°C |

What is the range?

A 16°C   **B** 21°C   **C** –27°C   **D** 27°C   **E** 13°C

50. 285.316

What is this number to 2 decimal places?

A 285.32   **B** 285.15   **C** 285.4   **D** 280   **E** 32

TEST ADVICE

This information will not appear in the actual test.
It is included here to remind you not to stop working
until you are told the test is over.

CHECK YOUR ANSWERS AGAIN IF THERE IS TIME

CORRECTING EVEN ONE MISTAKE CAN MEAN AN EXTRA MARK

Mathematics
Multiple-Choice
Practice Test C

Read the following carefully.

1. You must not open or turn over this booklet until you are told to do so.

2. This is a multiple-choice test, which contains a number of different types of questions.

3. You should do any rough working on a separate sheet of paper.

4. Answers should be marked in pencil on the answer sheet provided, not on the test booklet.

5. If you make a mistake, rub it out as completely as you can and put in your new answer.

6. Work as carefully and as quickly as you can. If you cannot do a question, do not waste time on it but go on to the next.

7. If you are not sure of an answer, choose the one you think is best.

8. You will have 50 minutes to complete the test.

1. The population of Frogstead is one and three quarter million. What is this number written in figures?

   A 103400   B 1050000   C 175000   D 1750000   E 1705000

2. The bar chart shows Year 8's favourite ice-cream flavours.

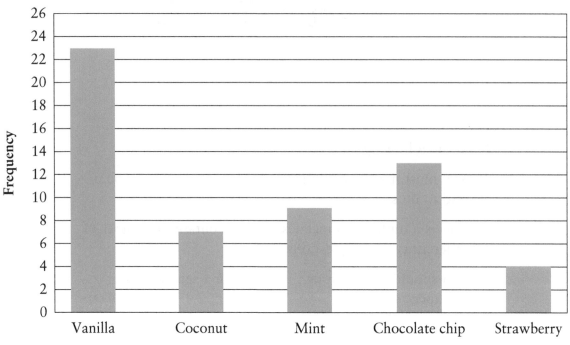

Ice-cream flavours

   How many children did not choose mint or coconut ice-cream?

   A 16   B 32   C 26   D 40   E 27

3. Eamon pays £7.85 per month for his mobile phone, 4p for each text message and 28p for each call. In March, he sends 37 text messages and makes 41 calls. How much is his phone bill in March?

   A £28.85   B £28.81   C £20.18   D £20.81   E £17.53

4. The weather forecast predicted that the coldest evening temperature in Archtown would be −6°. The forecast was not correct. It was actually 12° cooler than predicted. What was the actual temperature?

   A 12°   B −12°   C −7°   D 3°   E −18°

5. Akarsh, Rohan and Evan have measured out strings for conkers. Akarsh has a string with a length of 840 mm, Rohan has a string with a length of 0.07 m, Evan has a string with a length of 4.14 cm. What is the combined length of all their pieces of string?

   A 50300 mm   B 604 mm   C 283 mm   D 951.4 mm   E 855 mm

6. What is sixty-three tenths as a decimal?

   A 0.63   B 0.063   C 63.1   D 6.3   E 0.363

7. Gold University has 6000 students. 7.5% of the students own a bike. 13% of the students own a car. How many students own neither a bike nor a car?

   A 4770   B 5765   C 5795   D 4281   E 3666

8. Alba receives the marks from her end-of-term exams. The mark for her Chemistry exam is missing.

German	75%
English	82%
Maths	96%
Biology	79%
Chemistry	

   If her average mark is 83%, what percentage does she receive in Chemistry?

   A 76%   B 77%   C 68%   D 83%   E 85%

9. Which of these numbers is not a multiple of 35?

   A 175   B 210   C 435   D 245   E 3500

10. Look at this pictogram.

Number of cars	
Car park	Key: represents 26 cars
Parking A	(cars)
Parking B	(cars)
Parking C	(cars)

   How many more cars are parked in Parking C than in Parking B?

   A 78   B 84   C 91   D 92   E 85

11. Calculate 36% of 225

A 97   B 81   C 75   D 102   E 84

12. What is the 6th term in this sequence?

49, 64, 81, 100, …, ?

A 115   B 121   C 181   D 126   E 144

13. Mandy ran a 100 metre race and beat the previous school record of 13.83 seconds by thirty-six hundredths of a second. What was Mandy's time?

A 14.19 seconds   B 13.794 seconds   C 13.57 seconds   D 14.37 seconds
E 13.47 seconds

14. Lukas builds a path around his shed that is a uniform 2 m wide all the way round. The shed is 32 m by 6 m.

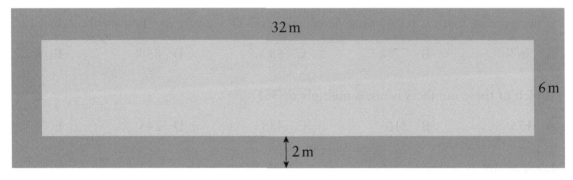

What is the total area of the path?

A 192 m²   B 178 m²   C 168 m²   D 204 m²   E 185 m²

15. Calculate 3794 divided by 7

A 542   B 649   C 487   D 546   E 586

16. 3784 people are in a stadium. 39 people leave and 241 are admitted. How many people are now in the stadium?

A 4216   B 3968   C 3986   D 2574   E 5012

17. Mrs Jawara wants to buy a shirt. The shirt usually costs £42.00 but is reduced by 18% in a sale. What is the sale price of the shirt?

A £7.56   B £30.24   C £34.44   D £39.76   E £16.54

18. Frank has 7 coins in his wallet. The total weight of the coins is 55.8 grams. If a 50 pence coin weighs 13.6 grams, a 1 pence coin weighs 3.5 grams and a 20 pence coin weighs 4 grams, how many 50 pence coins does he have?

    A 3   B 4   C 2   D 1   E 5

19. An apple orchard has 63 apple trees that each produce 793 apples. If Apple Farm collects all these apples and sells 36810 apples, how many apples are left over?

    A 13129   B 15256   C 13149   D 18302   E 25601

20. Winston makes a wooden crate. The height is 305 cm, the width is 540 cm and the length is 7.8 m.

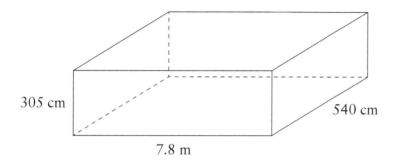

305 cm   540 cm   7.8 m

    What is the volume of his wooden crate in m³?

    A 1246.88 m³   B 128.664 m³   C 12847.45 m³   D 128.466 m³   E 12846600 m³

21. What is the name of this 3D shape?

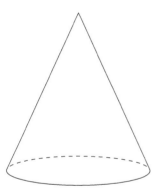

    A Sphere   B Cylinder   C Cuboid   D Triangular prism   E Cone

22. Which of these numbers is the greatest?

    A $\frac{5}{8}$ B $\frac{6}{9}$ C $\frac{4}{7}$ D 0.6   E $\frac{12}{15}$

23. Hot Chicken sauce is sold in bottles of 237 millilitres. Mr Williams has 8 bottles. In order to make his chicken marinade for 3 chickens, he must split the contents of the 8 bottles into 3. How many millilitres of sauce is required for each chicken?

    A 568 ml   B 634 ml   C 589 ml   D 632 ml   E 726 ml

24. Mira works in a bakery and measures out 79 kg of sugar per day. Which of these weights is closest to 79 kg?

   A 79.1 kg   **B** 78.899 kg   **C** 79.087 kg   **D** 78.9027 kg   **E** 78.909 kg

25. The distance between Pobbletown and Bobbletown is approximately $3\frac{3}{8}$ km. If Zorg walks $\frac{2}{3}$ of the way, how far does he walk?

   A $2\frac{7}{8}$ km   **B** $2\frac{5}{7}$ km   **C** $2\frac{1}{4}$ km   **D** $2\frac{1}{2}$ km   **E** $1\frac{7}{8}$ km

26. Which number lies halfway between 784 and 813?

   A 799.5   **B** 793   **C** 802.5   **D** 800.5   **E** 798.5

27. What is this number correct to 2 decimal places? 17.23571

   A 17.23   **B** 17.2   **C** 17.24   **D** 17.27   **E** 17.05

28. Here are the ingredients for a recipe to make 12 cookies.

   225 g butter
   110 g caster sugar
   275 g plain flour
   5 g mixed spice

   Tom wants to make 42 cookies. How much caster sugar will he need?

   A 305 g   **B** 290 g   **C** 285 g   **D** 385 g   **E** 310 g

29. Rose works 5 shifts, each lasting 7 hours 45 minutes. If she gets paid £7 per hour, how much will she have earned at the end of the 5 shifts?

   A £279.50   **B** £346.00   **C** £496.00   **D** £271.25   **E** £482.25

30. Calculate 9.1 divided by 0.7

   A 1.3   **B** 0.13   **C** 13   **D** 130   **E** 13.13

31. Here are the prices of 7 dining tables. What is the median price?

| £278 | £237 | £284 | £222 | £272 | £291 | £259 |

A £278   B £272   C £277   D £267.50   E £259

32. Max has a fair 6-sided spinner.

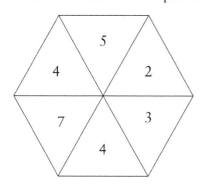

What is the probability of Max spinning a 7?

A $\frac{1}{6}$ B $\frac{1}{7}$ C $\frac{1}{8}$ D $\frac{7}{6}$ E $\frac{7}{1}$

33. Calculate 0.003 + 0.803 − 0.05

A 0.81   B 0.756   C 0.801   D 0.782   E 0.791

34. A snail truck travels 1740 m in 36 minutes. How far will the snail truck travel in 2 hours?

A 3480 m   B 4.57 km   C 5.25 km   D 5.8 km   E 6.1 km

35. Imran and Ramon have £184 which they share in the ratio 14:9. How much money does Ramon get?

A £16.56   B £87.25   C £86.75   D £72   E £207

36. A cookery programme lasts 85 minutes. If it starts at 7.38 p.m., what time will it finish?

A 22:05   B 21:03   C 21:33   D 20:54   E 22:48

37. Hamza has made 300 fruit pies. 171 are apple-flavoured and $\frac{4}{15}$ are plum-favoured and he keeps these for himself. The remainder are split between his siblings in the ratio 3:4. If Neil gets the bigger part, how many pies does he get?

A 46   B 56   C 49   D 35   E 28

38. Amelia draws a scalene triangle.

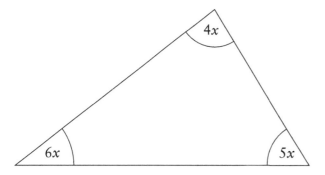

What is the value of *x*?

A 12°   B 23°   C 25°   D 10°   E 9°

39. Martin is buying a new house. The price of the house is three hundred and fifty-six thousand, and three pounds. What is this price written in figures?

A £35603   B £356030   C £356003   D £356303   E £365023

40. Pedro gets £1.65 pocket money every week. He is trying to save £23.10 for some football boots. How many weeks will it take him to save up enough money for the boots?

A 11   B 13   C 16   D 14   E 17

41. Sibel has a circular paddling pool. If the diameter is 247 cm, how far is it from the centre of the paddling pool to the edge?

A 247 cm   B 494 cm   C 124.5 cm   D 123.5 cm   E 207 cm

42. Wombley Stadium has 46820 seats for Wombles. On Thursday, 3714 seats are empty and the rest are filled. If tickets cost £25 each, how much will the stadium make on Thursday?

A £1077.65   B £10776.50   C £107765000   D £1077650   E £104765

43. If 8 apples cost £1, how much will it cost to buy 36?

A £4.25   B £4.35   C £4.50   D £4.65   E £3.50

44. What could you use to find the nth term of this sequence?

2, 10, 18, 26, 34, 42

A $2n + 8$ B $2(n + 8)$ C $2n + 8n$ D $8n - 6$ E $6n + 8$

45. 7 golf balls cost £5.60. How much would 26 golf balls cost?

A £20.80   B £22.40   C £25.30   D £27.45   E £18.65

46. Which one of these units would you use to measure the volume of a swimming pool?

    A Litres   **B** Kilograms   **C** Feet   **D** Cubic metres   **E** Pints

47. The scale shows the average weight of 25 oranges.

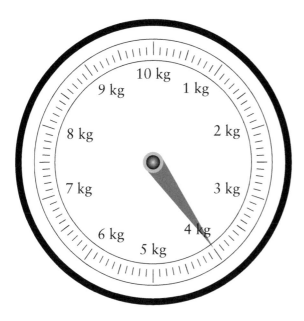

How much will 48 oranges weigh?

    A 7.68 kg   **B** 7.89 kg   **C** 7.56 kg   **D** 7.83 kg   **E** 7.7 kg

48. A bus travels at 53 kilometres per hour. How far will it have travelled after 4 hours?

    A 202 km   **B** 222 km   **C** 212 km   **D** 232 km   **E** 242 km

49. Sinita is 13 years younger than Anne. Write Anne's age (x) in relation to Sinita's age (y).

    A $x = y + 13$ **B** $x = 13y$ **C** $x = y - 13$ **D** $x = 2(y - 13)$ **E** $x = y - 3$

50. What are the total interior angles of 7 pentagons in degrees?

    A 2640°   **B** 3780°   **C** 3260°   **D** 3840°   **E** 2790°

TEST ADVICE

This information will not appear in the actual test.
It is included here to remind you not to stop working
until you are told the test is over.

CHECK YOUR ANSWERS AGAIN IF THERE IS TIME

CORRECTING EVEN ONE MISTAKE CAN MEAN AN EXTRA MARK

Mathematics
Multiple-Choice
Practice Test D

Read the following carefully.

1. You must not open or turn over this booklet until you are told to do so.

2. This is a multiple-choice test, which contains a number of different types of questions.

3. You should do any rough working on a separate sheet of paper.

4. Answers should be marked in pencil on the answer sheet provided, not on the test booklet.

5. If you make a mistake, rub it out as completely as you can and put in your new answer.

6. Work as carefully and as quickly as you can. If you cannot do a question, do not waste time on it but go on to the next.

7. If you are not sure of an answer, choose the one you think is best.

8. You will have 50 minutes to complete the test.

1. The population of Milton is eighty-six thousand and twenty-four. What is this number written in figures?

    A 8624   B 86240   C 860024   D 86024   E 80624

2. Kate stood outside her school counting the number of people in each passing car. The bar chart below shows the results.

    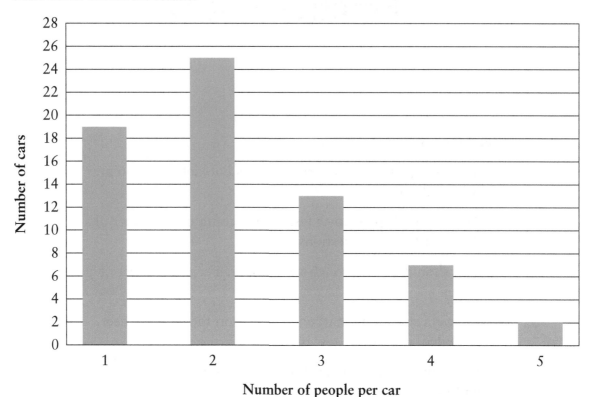

    How many cars contained more than 2 people?

    A 25   B 22   C 34   D 26   E 31

3. Nimishi looks at the clock. The time is twenty-five minutes to nine in the evening. What is this time written as a 24-hour clock time?

    A 9:25   B 21:25   C 19:35   D 22:35   E 20:35

4. Sunny School has 540 pupils. 60% of pupils have a dog. $\frac{1}{9}$ of pupils have a cat and 43 pupils have a fish. How many pupils don't have a pet?

    A 427   B 127   C 428   D 113   E 159

5. $72 \times 390 = 28080$

   What is 7200×0.00390?

   A 28.8   B 28.08   C 2880   D 28080   E 280800

6. Carmen is ill. The doctor has instructed her to take 4 tablets per day for 3 days. Each tablet weighs 375 mg. What is the total combined weight of the tablets she must take?

   A 0.45 g   B 4.5 g   C 3.05 g   D 1.005 g   E 5.4 g

7. A recipe for making 12 brownies includes the following ingredients. How many grams of butter will I need to make 60 brownies?

   > 225 g butter
   > 400 g brown sugar
   > 1 tablespoon vanilla extract
   > 4 eggs
   > 250 g plain flower
   > 140 g cocoa powder
   > $\frac{1}{4}$ teaspoon baking powder
   > 125 g mini chocolate chips

   A 1350 g   B 1250 g   C 675 g   D 900 g   E 1125 g

8. What is the 6th term in the sequence?

   0.16, 0.08, 0.04, 0.02, … , ?

   A 0.1   B 0.5   C 0.15   D 0.05   E 0.005

9. A box of chocolates weighs $\frac{1}{8}$ kg. The box contains 14 identical chocolates and the box itself weighs 13 g. Scott eats 5 chocolates. How many grams of chocolate does Scott eat?

   A 25 g   B 28 g   C 35 g   D 40 g   E 55 g

10. How many thousandths must be added to 6.026 to make 7?

    A 0.974   B 74   C 974   D 84   E 884

11. If 3rd May is a Thursday, what day of the week will 8th June be?

    A Friday   B Sunday   C Wednesday   D Monday   E Saturday

12. 450 teachers were asked which subject they taught at school. This pie chart shows the results.

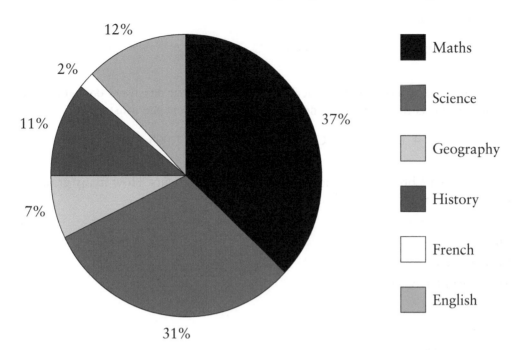

How many teachers did not teach science or maths?

A 306   **B** 32   **C** 144   **D** 59   **E** 45

13. Kwabena has 6 bags of sweets. The mean number of sweets in each bag is 7.
Kwabena removes 1 bag. The numbers of sweets left in the 5 remaining bags are:
8, 10, 3, 5, 7

How many sweets are in the bag that Kwabena removed?

A 11   **B** 4   **C** 7   **D** 9   **E** 8

14. Charlie and Min Yee play 2 sets of tennis. They play for a total of $1\frac{5}{6}$ hours. The 2nd set is 48 minutes longer than the 1st. What is the length of the 1st set?

A 17 minutes   **B** 23 minutes   **C** 36 minutes   **D** 28 minutes   **E** 31 minutes

15. Look at this graph.

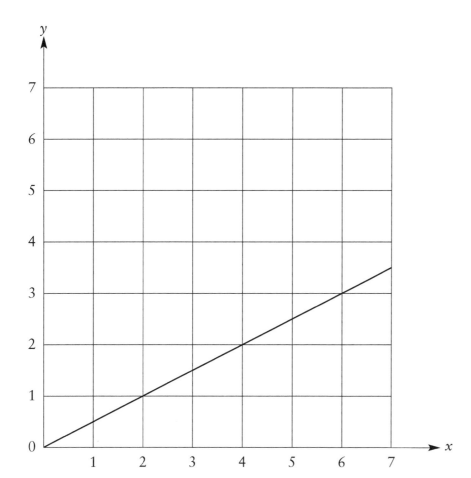

What is the equation of the line?

A $x = 2y$ B $x = y - 2$ C $x = 13$ D $x = 2$ E $x = y + 2$

16. Maya has a map. On the map, 7 mm represents 2.8 m. What is the scale of the map?

A 1:40   B 1:4   C 1:400   D 1:4000   E 1:7

17. Mr Lockey records the exact time it takes him to iron a single shirt on 5 consecutive days. His times are listed below.

Monday	151.04 seconds
Tuesday	163.1 seconds
Wednesday	150.2 seconds
Thursday	150.17 seconds
Friday	149.86 seconds

On which day was he closest to $2\frac{1}{2}$ minutes?

A Monday   B Tuesday   C Wednesday   D Thursday   E Friday

18. The train to Doncaster leaves at 19:48 travelling at 84 km per hour. How many kilometres has the train travelled when the time is 20:33?

A 40 km   B 44 km   C 50 km   D 66 km   E 63 km

19. Mr Brown packs pistachios into bags. Each bag contains 41 pistachios. How many bags can he fill from his stock of 697 pistachios?

A 24   B 23   C 17   D 11   E 15

20. Phoebe buys a sofa in the sale. The sofa's original price of £255 has been reduced by £108.73. What is the sale price of the sofa?

A £145.37   B £143.47   C £146.27   D £145.27   E £144.37

21. Farmer Jones uses $2b$ bales of hay per day to feed each of his adult horses and b bales of hay per day for each of his young horses. What expression would give the total number of hay bales he needs per day if he has 4 adult horses and 6 young horses?

A $4b + 6b$ B $12b$ C $6b + 8c$ D $14b$ E $10b$

22. What would you multiply 51235 by to get the answer 5.1235?

A 0.00001   B 100   C 0.001   D 1000   E 0.0001

23. How many eighths are there in 7?

A 8   B 7   C 15   D 80   E 56

24. The Adams family go to Rocku restaurant and order the following items from the menu below: 3 pizzas, 4 tomato soups, 4 drinks and 1 lasagne.

Tomato Soup	£1.85
Pizza	£4.99
Lasagne	£4.49
Drinks	75p

If they pay with a £50 note, how much change should they receive?

A £29.86   B £28.21   C £25.12   D £20.14   E £22.14

25. Neville has this money.

What is the total value of the money?

A £30.037   B £30.33   C £30.32   D £30.37   E £33.37

26. Ali is saving to buy a phone costing £185. He has already saved £27.50. He earns £7.50 per hour working in a café. How many hours must he work before he can afford to buy the phone?

A 23   B 21   C 16   D 19   E 25

27. Which of these expressions is equivalent to $3(n - 4) - 5$?

A $(3n - 4) - 5$ B $3n - 1$ C $3n - 7$ D $5 - (3n - 4)$ E $3n - 17$

28. Which of these numbers is exactly divisible by both 3 and 7?

A 203   B 567   C 391   D 526   E 489

29. Calculate 40% of 865

    A 302   **B** 324   **C** 362   **D** 354   **E** 346

30. Which of the following statements is not true?

    A A scalene triangle has no equal sides.

    B A rhombus has sides of equal length.

    C One pair of opposite sides in a trapezium are parallel.

    D A right-angled triangle has one angle of 90°.

    E A kite does not have two pairs of equal adjacent sides.

31. What percentage of the shape is shaded black?

    A 30%   **B** 40%   **C** 25%   **D** 35%   **E** 75%

32. There are 138 pencils in a box. Anwar is given 12 boxes. If he sells each pencil for 30p, how much can he expect to make if he sells every pencil?

    A £4140   **B** £414   **C** £4968   **D** £5214   **E** £496.80

33. The notice below shows the train timetable for Thamesbury to Bagelville. Part of the timetable is missing.

	Train 1	Train 2
Thamesbury	19:36	
Bagelville	21:05	23:41

If Train 1 and Train 2 take the same amount of time, what time must Brian catch the train from Thamesbury to arrive in Bagelville at 23:41?

    A 22:12   **B** 20:56   **C** 22:17   **D** 22:23   **E** 21:59

34. Lucia thinks of a number, doubles it and then subtracts 17. The answer is 19. What was the number Lucia first thought of?

 A 16   B 23   C 19   D 22   E 18

35. A hedgehog runs 17.5 metres in 5 seconds. If it continued at this exact speed, what would its average speed be in kilometres per hour?

 A 2.1 kph   B 12.6 kph   C 12600 kph   D 0.126 kph   E 0.85 kph

36. A baker sold 935 loaves of bread in 3 days. On the 1st day he sold 288 loaves. On the 2nd day he sold 150 more than on the 1st day. How many did he sell on the 3rd day?

 A 497   B 483   C 368   D 347   E 209

37. A shopping bag and its contents have a total mass of 3.075 kg. The bag weighs 20 g. The bag contains the following items:

3 bags of salad, each weighing 150 g

1 bag of carrots ($\frac{3}{4}$ kg)

5 chocolate bars, each weighing 35 g

4 packets of avocados

If 1 packet contains 3 avocados, how much does 1 avocado weigh?

 A 260 g   B 362 g   C 140 g   D 185 g   E 420 g

38. A square has the co-ordinates (1, 1), (4, 1), (1, 4), (4, 4). It is reflected in the x–axis. What are the new co-ordinates?

 A (4, 4), (1, 4), (1, 1), (–1, –4)

 B (4, –4), (1, 2), (4, –1), (2, –1)

 C (–1, –1), (–1, –4), (–4, –1), (–4, –4)

 D (1, –4), (–4, 3), (2, –4), (–3, –4)

 E (1, –1), (4, –1), (1, –4), (4, –4)

39. A piece of string has a length that is a whole number. It can be cut into both 15 cm and 9 cm equal pieces that are also a whole number. What is the shortest length that the string could be?

 A 135 cm   B 159 cm   C 45 cm   D 90 cm   E 30 cm

40. Look at the Venn diagram below and analyse why these numbers are in sets.

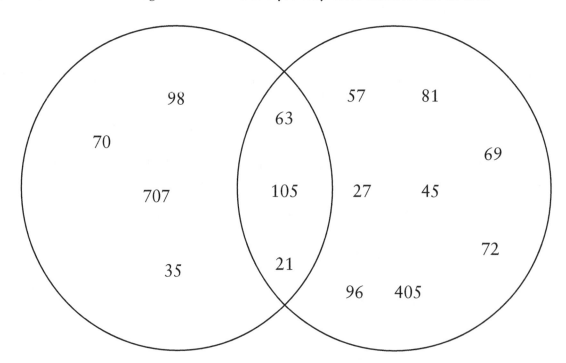

Which of these numbers could be in the overlapping section?

A 126   B 125   C 128   D 127   E 132

41. Madison, Isabella and Stella save £4.65 between them. Stella saves 25p more than Isabella. Madison saves 35p more than Isabella. How much does Isabella save?

A £1.25   B £1.40   C £1.45   D £1.35   E £1.55

42. A cockroach has a length of 45 mm. It walks 20 times its own length every day. How many days will it take the cockroach to walk 18 metres?

A 2   B 200   C 10   D 0.2   E 20

43. Farmer Faisal has a cabbage patch that is 13 metres by 7 metres. Farmer Bessie has a cabbage patch that has a perimeter of 102 metres. Farmer Bessie's cabbage patch is twice as long as it is wide. What is the difference between the areas of the 2 cabbage patches?

A 125 m²   B 487 m²   C 138 m²   D 403 m²   E 453 m²

44. What is $\frac{1}{4}$ of $\frac{2}{5}$ of 340?

A 38   B 46   C 50   D 34   E 44

45. The scale shows the weight of 12 peaches. If each peach weighs exactly the same, how much do 5 peaches weigh?

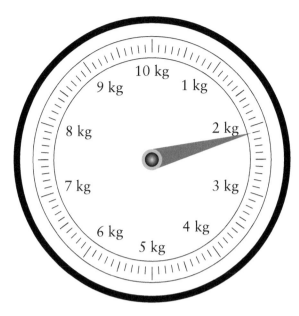

A 155 g   B 175 g   C 87.5 g   D 275 g   E 875 g

46. Ksenia sells pizzas and estimates that each pizza will take 5 minutes to heat up. It will take a further 2 minutes to serve the customer. What expression would give the total time in minutes if she serves *g* pizzas one at a time?

A 5*m* + 2*m* + *g*   B *g* + 7   C *g* – 7   D 5*g* + 2   E 7*g*

47. Which of these fractions is not equivalent to $\frac{3}{5}$?

A $\frac{33}{55}$ B $\frac{51}{85}$ C $\frac{72}{130}$ D $\frac{21}{35}$ E $\frac{300}{500}$

48. Robin has 5 books. Here are the weights of the 5 books.

126 g   320 g   55 g   138 g   347 g

What is the range?

A 126 g   B 138 g   C 209 g   D 292 g   E 314 g

49. What are the co-ordinates of points F and R?

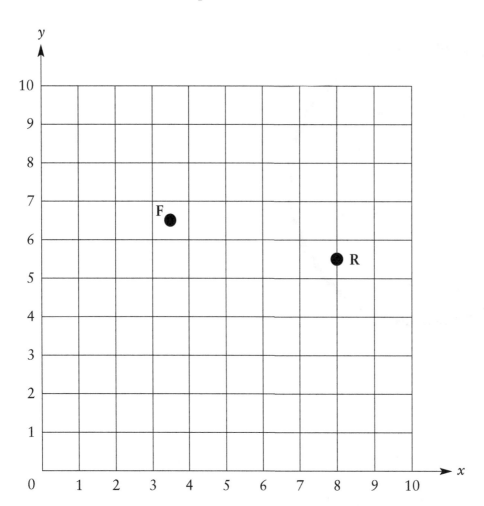

A (3, 3.5) and (8, 5) **B** (6, 3.5) and (5.5, 8) **C** (3.5, 6) and (7.5, 3.5)

D (3.5, 6.5) and (8, 5.5) **E** (4.5, 6) and (8, 6.5)

50. What percentage of this shape is shaded?

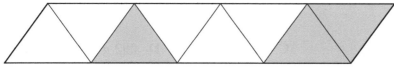

A 3% **B** 15% **C** 45% **D** 35.5% **E** 37.5%

Collins

PRACTICE PAPERS

Answers and Explanations

Mathematics

Practice Test A Answers and Explanations

1. **E 5640**
   2 hours = 120 minutes; number of words = 47 × 120 = 5640

2. **B 249**
   1492 ÷ 6 = 248.67; so 249 boxes are needed to fit all the eggs.

3. **E 37**
   $3x + 14 = 125$; $125 - 14 = 3x$; $3x = 111$; $x = 37$

4. **E 39**
   13% of 300 workers preferred doughnuts; 0.13 × 300 = 39

5. **C 54**
   Striker B scored x goals and Striker A scored $(x + 17)$ goals; $x + (x + 17) = 125$; $2x = 125 - 17$; $x = 108 ÷ 2 = 54$

6. **D 150**
   Bathroom wall = 200 cm × 300 cm = 60000 cm²; tiles = 20 × 20 = 400 cm²; 60000 ÷ 400 = 150

7. **D 1.5 hours**
   Total hours of sport = 19 + 5 + 1 + 13 + 4 + 3 = 45 hours. 30 days in September. Average time spent playing sport = 45 ÷ 30 = 1.5 hours

8. **B 100000**

9. **C 28**
   85% of 420 drive = 357; $\frac{1}{12}$ of 420 cycle = 35; number of employees who walk = 420 − 357 − 35 = 28

10. **B 87**
    There are 100 hundredths in 1 unit. 300 hundredths − 213 hundredths = 87 hundredths

11. **C £149.52**
    Perimeter = (13 × 2) + (15 × 2) = 56 m; wire fencing costs = £2.67 × 56 = £149.52

12. **C 20°**
    Angles on a straight line add up to 180°. $9y = 180°$; $y = 20°$

13. **D 48**
    Platform A = 6.5 × 16 = 104; Platform C = 3.5 × 16 = 56; Difference = 104 − 56 = 48 trains

14. **D 0.72**
    0.4 cm in 20 seconds = 1.2 cm in 60 seconds/1 minute; 1.2 cm × 60 = 72 cm in 1 hour = 0.72 metres per hour

15. **D 136**
    17 litres = 17000 ml; 17000 ÷ 125 = 136

16. **E 0.001**
    81430 × 0.001 = 81.43

17. **B £6.60**
    Sarah and Barry have 3 potatoes (3 × £1.50) and 7 toppings (7 × £0.30); £4.50 + £2.10 = £6.60

18. **B 7 weeks**
    2.548 kg = 2548 g. Roxy eats 26 × 2 = 52 g per day; 2548 ÷ 52 = 49 days = 7 weeks

19. **C 20**
    340 ÷ (85 × 0.2) = 340 ÷ 17 = 20

20. **E 33.3 m**
    Path = 1850 mm wide = 185 cm
    Perimeter = 2(700 + 185 + 185) + 2(225 + 185 + 185) = (2 × 1070) + (2 × 595) = 2140 + 1190 = 3330 cm = 33.3 m

21. **D 3x**
    Mr O'Murphy buys x tickets and Mr Bailey buys $2x$ tickets; $x + 2x = 3x$

22. **C 7**
    £6 = 600p; 600 ÷ 85 = 7 r 5

23. **E −26**
    The sequence pattern is −8, −9, −10, −11, −12, −13; so the sequence is 37, 29, 20, 10, −1, −13, −26

24. **C £300**
    387 ÷ 27 = 14 r 9 so 15 buses are required; 15 × £20 = £300

25. **E 4.92 and 5.08**

26. **A USA**

27. **A 3.996**
    4 − 3.996 = 0.004; 4.01 − 4 = 0.01; 4 − 3.99 = 0.01; 4.11 − 4 = 0.11; 4.007 − 4 = 0.007. Therefore 0.004 is the smallest difference so 3.996 is closest.

28. **D 1 hour 30 minutes**
    35 + 45 + 10 = 90 minutes = 1 hour 30 minutes

29. **E 184°**
    Carbohydrate content = 23 g. Each g = 360° ÷ 45 = 8°; 23 × 8 = 184°

30. **D 21:47**

31. **E 0.1%**
    £10 = 1000p; 1 ÷ 1000 × 100% = 0.1%

32. **C −23°C**
Tomorrow = −16 + 5 = −11°C; the day after tomorrow = −11°C − 12 = −23°C

33. **D 55 g**
Xan has measured 78 g. 78 − 23 = 55 g

34. **B 1394783**
1970530 − 575747 = 1394783

35. **E 476.5 g**
68p = 34 × 2p. Total weight = (34 × 7.25) + 230 = 246.5 + 230 = 476.5 g

36. **C 0**
Kirsty has x number of sweets, Stella has $2x$; $x = 45 \div 3 = 15$; Stella gets $2 \times 15 = 30$; 30 − 13 − 17 = 0

37. **C 172 minutes**
2 hours 52 minutes = 172 minutes

38. **C × 1000**
5601 × 1000 = 5601000

39. **B**

40. **D September**
£100 = 50p × 200 → 200 days. There are 31 days in March, May, July and August and 30 days in April, June and September.

41. **D 14**
Females under 50 = 39 − 21 = 18;
Males under 50 = 32 − 18 = 14

42. **B £14.28**
4 × £0.47 = £1.88; 12 × £0.32 = £3.84; £1.88 + £3.84 = £5.72; £20.00 − £5.72 = £14.28

43. **E Line V is perpendicular to line W.**

44. **D**
Shape D has 7 sides (a hexagon has 6 sides).

45. **A $5c − 4 = b$**
Adding 4 to both sides of the equation gives $5c + 4 = b$; so $5c − 4 = b$ cannot be true.

46. **A 45°**
Centre of octagon = circle; angles in the centre of a regular octagon = angles in a circle = 360°; 360° ÷ 8 = 45°

47. **D $4p + 24$**

48. **B (8, 9)**

49. **D 6100**

50. **E 4.48 cm**
35 × 1.28 = 44.8 mm = 4.48 cm

Practice Test B Answers and Explanations

1. **D £2.15**
   6 × 38p = 228p; 228p + 57p = 285p = £2.85. Change = £5 – £2.85 = £2.15

2. **C 21:38**
   116 minutes + 17 minutes = 2 hours 13 minutes. 7.25 p.m. + 2 hours 13 minutes = 9.38 p.m.

3. **B 2.003 kg**
   Remember that 1000 g = 1 kg
   2003 g = 2.003 kg

4. **D 31% of 105**
   $\frac{1}{4}$ of 112 = 28; $8\frac{1}{3}$; $\frac{5}{8}$ of 48 = (48 ÷ 8) × 5 = 30; 31% of 105 = 0.31 × 105 = 32.55; 20% of 145 = 0.2 × 145 = 29. Greatest value is 31% of 105

5. **E 240°**
   A complete circle is 360° and there are 12 numbers on a clock, so the angle between each number = 360° ÷ 12 = 30°. At 8 o'clock, one hand is pointing to 8 and one hand is pointing to 12. From 12 o'clock to 8 o'clock is 8 hours = 30° × 8 = 240°. Remember, a reflex angle is an angle that is greater than 180° and less than 360°.

6. **C 27400**
   (43 + 57) × 274 = 100 × 274 = 27400

7. **D 35g**
   Total weight of hamper = 1.43 kg = 1430 g; 1430 – 450 = 980 g; 980 ÷ 28 = 35. Each chocolate bar weighs 35 g.

8. **B 52°C**
   Mumbai is the hottest city (35°C). Moscow is the coldest city (–17°C). The difference in temperatures = 35 – –17 = 35 + 17 = 52°C

9. **D 38**
   Let the number be *n*. (3*n* – 50) ÷ 4 = 16; 3*n* – 50 = 16 × 4 = 64. 3*n* = 64 + 50 = 114; *n* = 114 ÷ 3 = 38

10. **E £2.60**
    $\frac{1}{4}$ of 284 = 71; 284 – 71 = 213; $\frac{4}{6}$ of remaining amount = (213 ÷ 6) × 4 = 142. Amount to be sold = 213 – 142 – 6 = 65. Money made = 65 × 4p = 260p = £2.60

11. **B 45**
    Mean = average = total sandwiches ÷ number of days; (48 + 49 + 41 + 24 + 37 + 50 + 66) ÷ 7 = 315 ÷ 7 = 45 sandwiches per day

12. **D 20%**
    Retail price = £2.95; Hot chocolate costs = 37p + ($\frac{1}{4}$ × 88p) = 37p + 22p = 59p. $\frac{59}{295}$ = 0.2 = 20%

13. **C 450 cm**
    The area of patch A = 12 × 3 = 36 m². The area of patch B = 36 m² so the width is 36 ÷ 8 = 4.5 m = 450 cm.

14. **A 2.9962**

15. **E 228**
    Eldest child is Ronald. He was born on 30th November 1991. Youngest child is Daniel. He was born on 15th July 1992. Difference = 31 days (December) + 31 days (January) + 29 days (February) + 31 days (March) + 30 days (April) + 31 days (May) + 30 days (June) + 15 days (July) = 228 days.

16. **C Rhombus**
    A rhombus is a quadrilateral (has 4 sides), with all sides equal and opposite angles equal.

17. **E £3.50**
    Mrs Jules parks from 9.45 to 18.05 = 8 hours 20 minutes. Cost for 8 hours = 42p × 8 = 336p. 20 minutes = $\frac{1}{3}$ of an hour. Cost for 20 minutes = 42 ÷ 3 = 14p. Total cost = 336p + 14p = 350p = £3.50

18. **D £19.04**
    Total area of the tiles = 3.5 × 4 = 14 m². Cost = £1.36 × 14 = £19.04

19. **C 57°**
    Total of angles in a triangle = 180°. *x* = 180 – 49 – 74 = 180 – 123 = 57°

20. **D 378.04**
    One hundredth = 0.01. 378.036 to the nearest 0.01 = 378.04

21. **C 273**
    Ratio of boys:girls = 17:4. If 52 are girls then each part = 52 ÷ 4 = 13. There are 13 × 17 boys = 221. So in total there are 52 + 221 = 273 pupils.

22. **A 84 m**
    1 cm represents 7 m. So width = 2.5 × 7 = 17.5 m. Length = 3.5 × 7 = 24.5 m Perimeter = 2 × width (2 × 17.5) + 2 × length (2 × 24.5) = 35 + 49 = 84 m

23. **D 18**

   105 candles ÷ 6 = 17.5 so 18 boxes must be used to fit all the candles in.

24. **E 0.0342 km**

   1 cm represents 3420 cm = 34.2 m (100 cm in 1 m). 34.2 m = 0.0342 km Remember that 1000 m = 1 km

25. **C $2\frac{1}{2}$**

   $7\frac{2}{5}$ = 7.4; $4\frac{9}{10}$ = 4.9. So the difference (how much bigger) = 7.4 − 4.9 = 2.5 = $2\frac{1}{2}$

26. **A 7**

   £2.68 × 7 = £18.76. 8 would be too many because £2.68 × 8 = £21.44 and Jamil has £20.

27. **C 857**

   $\frac{4}{5}$ of 950 = (950 ÷ 5) × 4 = 760; $\frac{9}{10}$ of 950 = (950 ÷ 10) × 9 = 855; 857; 0.75 of 950 = 0.75 × 950 = 712.5; $\frac{1}{2}$ of 950 = 475. Greatest value = 857

28. **E 182000 m**

   Distance travelled in 3 hours = 3 × 56 = 168 km. 15 minutes = $\frac{1}{4}$ of an hour. Distance travelled in $\frac{1}{4}$ hour = 56 ÷ 4 = 14. So total distance travelled = 168 km + 14 km = 182 km = 182000 m

29. **A 2703 m**

   Salim walks 3.05 km per day = 3050 m; Salim's father walks 3050 − 347 = 2703 m

30. **D 62**

   The sequence is −9 each time, therefore as the 4th term is 80, the 6th term is 62.

31. **B 102 km**

   17 km/15 minutes = 68 km/60 minutes = 68 km per hour. Distance travelled = 68 × 1.5 = 102 km

32. **E 45**

   There are 5 fifths in 1; 9 × 5 = 45

33. **D 36**

   15% of 240 preferred science fiction. 10% = 24 and 5% = 12; 15% = 36

34. **E 34793**

   There are 1000 thousandths in 1 unit.

35. **B 56p**

   1 roll costs 315 ÷ 45 = 7p; 2 rolls = 2 × 7 = 14p; 1 slice of ham costs 350 ÷ 25 = 14p; 3 slices of ham cost 3 × 14 = 42p; 14p + 42p = 56p

36. **E $y = 3x$**

   y = 3 and x = 1, so $y = 3x$.

37. **D $14\frac{3}{4}$**

38. **D 57**

   Hammerville scored 61 goals in February and 4 in January. 61 − 4 = 57

39. **A Wednesday**

   There are 31 days in March. 15 days to the end of March + 11 days in April = 26 days. There are 7 days in the week and so 26 days later will be a Wednesday.

40. **D 1:40000**

   50 cm represents 20 km; scale = 50 cm:2000000 cm = 1:40000. Remember that 1 km = 100000 cm

41. **B 625**

   10 euros = approx. 250 yen; therefore 25 euros = 2.5 × 250 yen = 625 yen

42. **D 268.5 m²**

   Treat the garden patch as two rectangles. 3 × 8.5 = 25.5 m² and 9 × 27 = 243 m²; total area = 25.5 + 243 = 268.5 m²

43. **D 43.1 kg**

   10 g = 0.01 kg, so 43.097 rounded to 2 d.p. = 43.10 = 43.1 kg

44. **D 434 g**

   Weight of 1 pizza slice = 1668 g ÷ 6 = 278 g; total weight of pizza slice = 278 + (12 × 13) = 278 + 156 = 434 g

45. **C 96**

   $\frac{1}{3}$ of 432 = 144; $\frac{4}{6}$ of 144 = (144 ÷ 3) × 2 = 96

46. **D £5.97**

   1 ice-cream = 325 ÷ 25 = 13p; 31 × 13p = 403p = £4.03; change from £10.00 = £10.00 − £4.03 = £5.97

47. **E 161**

   10% of 700 = 70; 20% of 700 = 140; 1% of 700 = 7; 3% of 700 = 21; 23% of 700 = 140 + 21 = 161

48. **D 3**

   Total of numbers = 5 × 8 = 40. 40 − 8 − 13 − 13 = 6. The other 2 numbers must add up to 6 but are the same → 6 ÷ 2 = 3

49. **D 27°C**

   Range = highest temperature − lowest temperature = 12 − −15 = 12 + 15 = 27°C

50. **A 285.32**

Practice Test C Answers and Explanations

1. **D 1750000**
   Three-quarters of a million is 750000.
2. **D 40**
   23 (Vanilla) + 13 (Chocolate chip) + 4 (Strawberry) = 40
3. **D £20.81**
   £7.85 = 785p; 785p + (37 × 4p) + (41 × 28p) = 785p + 148p + 1148p = 2081p = £20.81
4. **E –18°**
   –6° – 12° = –18°
5. **D 951.4 mm**
   Combined length = 840 mm + 70 mm + 41.4 mm = 951.4 mm
6. **D 6.3**
7. **A 4770**
   7.5% of 6000 = 450; 0.13 × 6000 = 780; students with a bike or car = 450 + 780 = 1230. Students with no bike or car = 6000 – 1230 = 4770
8. **D 83%**
   Average mark = 83; total marks = 83 × 5 = 415; Chemistry mark = 415 – (75 + 82 + 96 + 79) = 83%
9. **C 435**
   435 ÷ 35 = 12.428... 12.428... is not a whole number so 435 is not a multiple of 35.
10. **C 91**
    Parking C = 5.5 × 26 = 143 cars; Parking B = 2 × 26 = 52 cars. Difference = 143 – 52 = 91
11. **B 81**
    0.36 × 225 = 81
12. **E 144**
    The sequence is 7^2, 8^2, 9^2, 10^2 OR +15, +17, +19, +21 (adding 2 more each time); so the 6th term in the sequence = 12^2 = 144
13. **E 13.47 seconds**
    Thirty-six hundredths = 0.36. Mandy ran a faster time to beat the previous school record. Mandy's time = 13.83 – 0.36 = 13.47 seconds.
14. **C 168 m²**
    Total area = (32 + 2 + 2) × (6 + 2 + 2) = 36 × 10 = 360 m²; area of shed = 32 × 6 = 192 m²; area of path = 360 – 192 = 168 m²
15. **A 542**
16. **C 3986**
    3784 – 39 + 241 = 3986
17. **C £34.44**
    Sale price = 100% – 18% of original = 82%; 82% of £42 = $\frac{82}{100}$ × 42 = £34.44
18. **A 3**
    13.6 needs to be multiplied enough times to end in 0.8; 3 × 13.6 = 40.8
19. **C 13149**
    63 × 793 = 49959 apples; 49959 – 36810 = 13149
20. **D 128.466 m³**
    Crate volume = 3.05 × 5.4 × 7.8 = 128.466 m³
21. **E Cone**
22. **E $\frac{12}{15}$**
    The easiest method is to convert all fractions to decimals. $\frac{5}{8}$ = 0.625; $\frac{6}{9}$ = 0.666...; $\frac{4}{7}$ = 0.571...; 0.6; $\frac{12}{15}$ = 0.8. So $\frac{12}{15}$ is the greatest.
23. **D 632 ml**
    237 × 8 = 1896 ml; 1896 ÷ 3 = 632 ml
24. **C 79.087 kg**
    Calculate the difference for each option. 79.1 – 79 = 0.1; 79 – 78.899 = 0.101; 79.087 – 79 = 0.087; 79 – 78.9027 = 0.0973; 79 – 78.909 = 0.091; 0.087 is the smallest difference so 79.087 kg is the closest to 79 kg.
25. **C $2\frac{1}{4}$ km**
    $3\frac{3}{8}$ km = $\frac{27}{8}$ km; $\frac{27}{8}$ km × $\frac{2}{3}$ = $\frac{54}{24}$ = $\frac{9}{4}$ = $2\frac{1}{4}$ km
26. **E 798.5**
    Halfway = (784 + 813) ÷ 2 = 798.5
27. **C 17.24**
28. **D 385 g**
    12 cookies requires 110 g caster sugar; 42 cookies requires (42 ÷ 12) × 110 g = 385 g
29. **D £271.25**
    5 × 7 hours 45 minutes = 5 × 7.75 hours = 38.75 hours; 38.75 × £7 = £271.25
30. **C 13**
    9.1 ÷ 0.7 = 13

31. **B £272**
    Median price = middle price. Arranged in order, the prices are £222, £237, £259, £272, £278, £284, £291. So the middle price is £272.

32. **A $\frac{1}{6}$**
    There is only 1 side with the number 7 so the probability = $\frac{1}{6}$

33. **B 0.756**

34. **D 5.8 km**
    Number of minutes in 2 hours = 2 × 60 = 120; distance travelled in 120 minutes = (120 ÷ 36) × 1740 = 5800 m = 5.8 km

35. **D £72**
    Number of parts = 14 + 9 = 23; value of each part = £184 ÷ 23 = £8; Ramon has 9 parts so he gets £8 × 9 = £72

36. **B 21:03**
    85 minutes = 1 hour 25 minutes; 7.38 p.m. + 1 hour + 25 minutes = 9.03 p.m. = 21:03

37. **E 28**
    $300 - 171 - (\frac{4}{15} \times 300) = 300 - 171 - 80 = 49$; the remainder are split into 7 parts so 1 part = 49 ÷ 7 = 7; Neil gets 4 × 7 = 28

38. **A 12°**
    Sum of internal angles in a triangle = 180°; $(4x + 5x + 6x) = 15x = 180°$; $x = 180° ÷ 15 = 12°$

39. **C £356003**

40. **D 14**
    £23.10 ÷ £1.65 = 14

41. **D 123.5 cm**
    Distance from the centre of the paddling pool to the circumference = diameter ÷ 2 = 247 ÷ 2 = 123.5 cm

42. **D £1077650**
    46820 – 3714 = 43106 seats; 43106 × £25 = £1077650

43. **C £4.50**
    8 apples cost £1; 36 ÷ 8 = 4.5; so 36 apples cost £1 × 4.5 = £4.50

44. **D $8n - 6$**
    The sequence increases by 8 each time so the nth term must include $8n$. The 1st term is 2 which is $8(1) - 6$. So the nth term is $8n - 6$.

45. **A £20.80**
    Cost of 1 golf ball = £5.60 ÷ 7 = £0.80; cost of 26 golf balls = £0.80 × 26 = £20.80

46. **D Cubic metres**

47. **A 7.68 kg**
    Weight of 25 oranges = 4 kg; weight of 1 orange = 4 kg ÷ 25 = 0.16 kg; weight of 48 oranges = 48 × 0.16 kg = 7.68 kg

48. **C 212 km**
    53 × 4 = 212 km

49. **A $x = y + 13$**

50. **B 3780°**
    Interior angles in one pentagon = 540°; in seven pentagons = 540° × 7 = 3780°

Practice Test D Answers and Explanations

1. D 86024
2. B 22
   13 (3 people) + 7 (4 people) + 2 (5 people) = 22
3. E 20:35
4. D 113
   Pupils with a dog = 60% of 540 = 0.6 × 540 = 324; pupils with a cat = $\frac{1}{9}$ × 540 = 60; pupils with a pet = 324 + 60 + 43 = 427; pupils without a pet = 540 − 427 = 113
5. B 28.08
6. B 4.5g
   4 tablets × 3 days = 12 tablets; 375 × 12 = 4500 mg = 4.5 g. Remember that 1g = 1000 mg.
7. E 1125g
   Grams of butter for 60 brownies = (60 ÷ 12) × 225 g = 5 × 225 g = 1125 g
8. E 0.005
   The sequence is half of the previous term each time. 5th term = 0.02 ÷ 2 = 0.01; 6th term = 0.01 ÷ 2 = 0.005
9. D 40 g
   Box of chocolates = $\frac{1}{8}$ kg = 125 g; weight of 14 chocolates = 125 − 13 = 112 g; weight of one chocolate = 112 ÷ 14 = 8 g; weight of chocolate eaten by Scott = 5 × 8 = 40 g
10. C 974
    7 − 6.026 = 0.974; number of thousandths = 974
11. A Friday
    There are 31 days in May. Number of days between 3rd May and 8th June = days remaining in May (31 − 3) + days in June (8) = 31 − 3 + 8 = 36 days. Number of full weeks in 36 days = 36 ÷ 7 = 5 remainder 1. Day of the week = Thursday + 1 day = Friday.
12. C 144
    Percentage of teachers = 100 − 31 − 37 = 32%; 32% of 450 = 0.32 × 450 = 144
13. D 9
    Total number of sweets = 7 × 6 = 42 Number of sweets in bag = 42 − 8 − 10 − 3 − 5 − 7 = 9

14. E 31 minutes
    $1\frac{5}{6}$ hours = 1 hour 50 minutes = 110 minutes. Let x = length of 1st set: $x + (x + 48)$ = 110; $2x + 48 = 110$; $2x = 62$; x = 31 minutes
15. A $x = 2y$
    The line passes through (4, 2); $x = 4$, $y = 2$; $x = \frac{4}{2} y$; $x = 2y$
16. C 1:400
    7 mm:2.8 m = 7 mm:2800 mm = 1:400
17. E Friday
    2.5 minutes = 150 seconds. Monday difference is 1.04 seconds; Tuesday difference is 13.1 seconds; Wednesday difference is 0.2 seconds; Thursday difference is 0.17 seconds; Friday difference is 0.14 seconds; Friday is the closest.
18. E 63 km
    19:48 to 20:33 is 45 minutes = $\frac{3}{4}$ of an hour. Distance = (84 ÷ 4) × 3 = 63 km
19. C 17
    Number of bags = 697 ÷ 41 = 17
20. C £146.27
    £255 − £108.73 = £146.27
21. D 14b
    $(4 \times 2b) + (6 \times b) = 8b + 6b = 14b$
22. E 0.0001
    51235 × 0.0001 = 5.1235
23. E 56
    There are 8 eighths in 1. Eighths in 7 = 8 × 7 = 56
24. D £20.14
    Cost of meal = 3 pizzas (3 × £4.99) + 4 soups (4 × £1.85) + 4 drinks (4 × £0.75) + 1 lasagne (£4.49) = £14.97 + £7.40 + £3 + £4.49 = £29.86; change = £50.00 − £29.86 = £20.14
25. D £30.37
    (3 × £10) + (3 × 10p) + (2 × 2p) + (3 × 1p) = £30 + £0.30 + £0.04 + £0.03 = £30.37
26. B 21 hours
    Money still needed = £185 − £27.50 = £157.50; hours to work = 157.50 ÷ 7.50 = 21 hours

27. **E** $3n - 17$

$3(n - 4) - 5 = 3n - 12 - 5 = 3n - 17$
Remember that when multiplying expressions in brackets, everything in the brackets must be multiplied by the number outside the bracket, so $3(n) = 3n$, and $3(4) = 3 \times 4 = 12$

28. **B** 567

$567 \div 3 = 189$ and $567 \div 7 = 81$

29. **E** 346

$10\% = 86.5$; $40\% = 86.5 \times 4 = 346$

30. **E A kite does not have two pairs of equal adjacent sides.**

31. **E** 75%

Work out how many rows and columns of smaller rectangles there are. Total number of rectangles = 8 (long side) × 5 (short side) = 40. There are 10 white rectangles. Number of shaded rectangles = 40 − 10 = 30. Percentage of shape that is shaded = $(30 \div 40) \times 100$ = 75%

32. **E** £496.80

Number of pencils = $138 \times 12 = 1656$; money made = $1656 \times £0.30 = £496.80$

33. **A** 22:12

Journey time = 1 hour 29 minutes. Brian must catch Train 2 1 hour and 29 minutes before 23:41 = 22:12

34. **E** 18

Let the number be x. $2x - 17 = 19$; $2x = 19 + 17 = 36$; $x = 18$

35. **B** 12.6 kph

Distance travelled in 1 minute = $17.5 \times 12 = 210$ metres; distance travelled in 1 hour = $210 \times 60 = 12600$ metres. Speed = 12.6 kph

36. **E** 209

$935 - 288 - (288 + 150) = 209$

37. **C** 140 g

3.075 kg = 3075 g. Weight of 4 avocado packets = 3075 g − 20 g − (3 × 150 g) − 750 g − (5 × 35 g) = 3075 − 20 − 450 − 750 − 175 = 1680 g; weight of 1 avocado packet = 1680 ÷ 4 = 420 g; weight of 1 avocado = 420 ÷ 3 = 140 g

38. **E**

Reflecting the x axis moves all the points to the same distance below the x axis.

39. **C** 45 cm

$45 \div 15 = 3$; $45 \div 9 = 5$; 45 is the smallest number that can be divided exactly by both 15 and 9.

40. **A** 126

The numbers in the left-hand circle are all divisible by 7; the numbers in the right-hand circle are all divisible by 3.

41. **D** £1.35

Let x be the amount that Isabella saves. £4.65 = 465p; $x + (x + 35p) + (x + 25p) = 465p$; $3x + 60p = 465p$; $3x = 465p - 60p = 405p$; $x = 405p \div 3 = 135p = £1.35$

42. **E** 20

Distance walked in 1 day = 45 mm × 20 = 900 mm = 90 cm; 18 m = 1800 cm; days to walk 1800 cm = 1800 ÷ 90 = 20

43. **B** 487 m²

Area of Faisal's cabbage patch = $13 \times 7 = 91$ m². For Bessie's cabbage patch, let x be the width: $x + 2x + x + 2x = 102$ m; $6x = 102$; $x = 17$ m. Length = 2 × 17 = 34 m. Area of Bessie's cabbage patch = $17 \times 34 = 578$ m²; difference between areas = $578 - 91 = 487$ m²

44. **D** 34

$\frac{2}{5}$ of 340 = 136; $\frac{1}{4}$ of 136 = 34

45. **E** 875 g

Weight of 1 peach = 2.1 ÷ 12 = 0.175 kg = 175 g; weight of 5 peaches = 5 × 175 = 875 g

46. **E** $7g$

Time to serve one pizza = 5 + 2 = 7; time to serve g (number of) pizzas = $7 \times g = 7g$

47. **C** $\frac{72}{130}$

48. **D** 292 g

Range = 347 g − 55 g = 292 g

49. **D** (3.5, 6.5) and (8, 5.5)

50. **E** 37.5%

$\frac{3}{8}$ is shaded. Percentage = $(3 \div 8) \times 100$ = 37.5%

Notes

Notes

Notes

Notes

Notes

Pupil's Name

School Name

Date of Test

PUPIL NUMBER

[0]	[0]	[0]	[0]	[0]	[0]
[1]	[1]	[1]	[1]	[1]	[1]
[2]	[2]	[2]	[2]	[2]	[2]
[3]	[3]	[3]	[3]	[3]	[3]
[4]	[4]	[4]	[4]	[4]	[4]
[5]	[5]	[5]	[5]	[5]	[5]
[6]	[6]	[6]	[6]	[6]	[6]
[7]	[7]	[7]	[7]	[7]	[7]
[8]	[8]	[8]	[8]	[8]	[8]
[9]	[9]	[9]	[9]	[9]	[9]

SCHOOL NUMBER

[0]	[0]	[0]	[0]	[0]	[0]	[0]
[1]	[1]	[1]	[1]	[1]	[1]	[1]
[2]	[2]	[2]	[2]	[2]	[2]	[2]
[3]	[3]	[3]	[3]	[3]	[3]	[3]
[4]	[4]	[4]	[4]	[4]	[4]	[4]
[5]	[5]	[5]	[5]	[5]	[5]	[5]
[6]	[6]	[6]	[6]	[6]	[6]	[6]
[7]	[7]	[7]	[7]	[7]	[7]	[7]
[8]	[8]	[8]	[8]	[8]	[8]	[8]
[9]	[9]	[9]	[9]	[9]	[9]	[9]

DATE OF BIRTH

Day		Month		Year	
[0]	[0]	January	▭	2007	▭
[1]	[1]	February	▭	2008	▭
[2]	[2]	March	▭	2009	▭
[3]	[3]	April	▭	2010	▭
	[4]	May	▭	2011	▭
	[5]	June	▭	2012	▭
	[6]	July	▭	2013	▭
	[7]	August	▭	2014	▭
	[8]	September	▭	2015	▭
	[9]	October	▭	2016	▭
		November	▭	2017	▭
		December	▭	2018	▭

1
- 94 ▭
- 9400 ▭
- 6674 ▭
- 5880 ▭
- 5640 ▭

2
- 237 ▭
- 249 ▭
- 252 ▭
- 248 ▭
- 244 ▭

3
- 39 ▭
- 40 ▭
- 43 ▭
- 36 ▭
- 37 ▭

4
- 130 ▭
- 13 ▭
- 25 ▭
- 37 ▭
- 39 ▭

5
- 51 ▭
- 57 ▭
- 54 ▭
- 52 ▭
- 55 ▭

6
- 20 ▭
- 60 ▭
- 160 ▭
- 150 ▭
- 80 ▭

7
- 0.5 h ▭
- 3 h ▭
- 2.25 h ▭
- 1.5 h ▭
- 2.5 h ▭

8
- 1000 ▭
- 100000 ▭
- 10000 ▭
- 15000 ▭
- 1000000 ▭

9
- 36 ▭
- 24 ▭
- 28 ▭
- 48 ▭
- 32 ▭

10
- 970 ▭
- 87 ▭
- 0.97 ▭
- 0.87 ▭
- 870 ▭

11
- £40.05 ▭
- £520.65 ▭
- £149.52 ▭
- £187.69 ▭
- £231.58 ▭

12
- 15° ▭
- 10° ▭
- 20° ▭
- 180° ▭
- 120° ▭

13
- 56 ▭
- 44 ▭
- 58 ▭
- 48 ▭
- 62 ▭

14
- 1.2 ▭
- 7.2 ▭
- 72 ▭
- 0.72 ▭
- 12 ▭

15
- 102 ▭
- 215 ▭
- 137 ▭
- 136 ▭
- 224 ▭

16
- 1000 ▭
- 0.01 ▭
- 0.0001 ▭
- 100 ▭
- 0.001 ▭

17
- £4.50 ▭
- £6.60 ▭
- £5.80 ▭
- £5.85 ▭
- £6.70 ▭

18
- 4 weeks ▭
- 7 weeks ▭
- 5 weeks ▭
- 9 weeks ▭
- 10 weeks ▭

19
- 2 ▭
- 4 ▭
- 20 ▭
- 8 ▭
- 12 ▭

20
- 33.03 m ▭
- 32.15 m ▭
- 28.65 m ▭
- 29.5 m ▭
- 33.3 m ▭

21
- $x + 2y$ ▭
- $x + 2$ ▭
- $x - 2$ ▭
- $3x$ ▭
- $3x - 1$ ▭

22
- 6 ▭
- 5 ▭
- 7 ▭
- 9 ▭
- 8 ▭

23
- −13 ▭
- −18 ▭
- 24 ▭
- 26 ▭
- −26 ▭

24
- £240 ▭
- £320 ▭
- £300 ▭
- £280 ▭
- £340 ▭

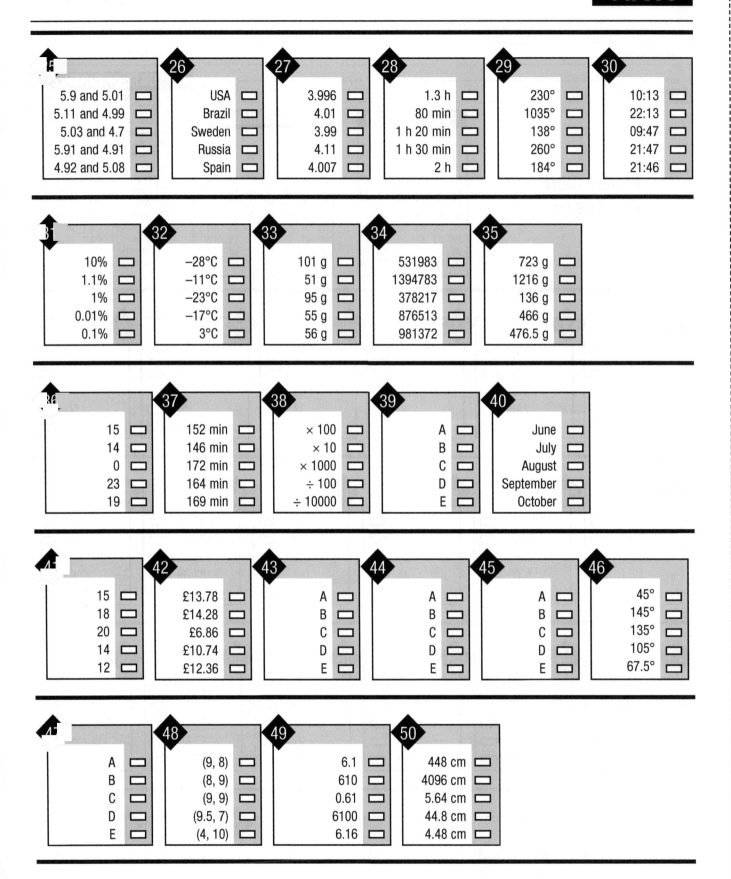

25
- 5.9 and 5.01 ☐
- 5.11 and 4.99 ☐
- 5.03 and 4.7 ☐
- 5.91 and 4.91 ☐
- 4.92 and 5.08 ☐

26
- USA ☐
- Brazil ☐
- Sweden ☐
- Russia ☐
- Spain ☐

27
- 3.996 ☐
- 4.01 ☐
- 3.99 ☐
- 4.11 ☐
- 4.007 ☐

28
- 1.3 h ☐
- 80 min ☐
- 1 h 20 min ☐
- 1 h 30 min ☐
- 2 h ☐

29
- 230° ☐
- 1035° ☐
- 138° ☐
- 260° ☐
- 184° ☐

30
- 10:13 ☐
- 22:13 ☐
- 09:47 ☐
- 21:47 ☐
- 21:46 ☐

31
- 10% ☐
- 1.1% ☐
- 1% ☐
- 0.01% ☐
- 0.1% ☐

32
- −28°C ☐
- −11°C ☐
- −23°C ☐
- −17°C ☐
- 3°C ☐

33
- 101 g ☐
- 51 g ☐
- 95 g ☐
- 55 g ☐
- 56 g ☐

34
- 531983 ☐
- 1394783 ☐
- 378217 ☐
- 876513 ☐
- 981372 ☐

35
- 723 g ☐
- 1216 g ☐
- 136 g ☐
- 466 g ☐
- 476.5 g ☐

36
- 15 ☐
- 14 ☐
- 0 ☐
- 23 ☐
- 19 ☐

37
- 152 min ☐
- 146 min ☐
- 172 min ☐
- 164 min ☐
- 169 min ☐

38
- × 100 ☐
- × 10 ☐
- × 1000 ☐
- ÷ 100 ☐
- ÷ 10000 ☐

39
- A ☐
- B ☐
- C ☐
- D ☐
- E ☐

40
- June ☐
- July ☐
- August ☐
- September ☐
- October ☐

41
- 15 ☐
- 18 ☐
- 20 ☐
- 14 ☐
- 12 ☐

42
- £13.78 ☐
- £14.28 ☐
- £6.86 ☐
- £10.74 ☐
- £12.36 ☐

43
- A ☐
- B ☐
- C ☐
- D ☐
- E ☐

44
- A ☐
- B ☐
- C ☐
- D ☐
- E ☐

45
- A ☐
- B ☐
- C ☐
- D ☐
- E ☐

46
- 45° ☐
- 145° ☐
- 135° ☐
- 105° ☐
- 67.5° ☐

47
- A ☐
- B ☐
- C ☐
- D ☐
- E ☐

48
- (9, 8) ☐
- (8, 9) ☐
- (9, 9) ☐
- (9.5, 7) ☐
- (4, 10) ☐

49
- 6.1 ☐
- 610 ☐
- 0.61 ☐
- 6100 ☐
- 6.16 ☐

50
- 448 cm ☐
- 4096 cm ☐
- 5.64 cm ☐
- 44.8 cm ☐
- 4.48 cm ☐

MA B

Pupil's Name

School Name

Date of Test

DATE OF BIRTH		
Day	Month	Year

Day		Month		Year	
[0]	[0]	January ☐		2007 ☐	
[1]	[1]	February ☐		2008 ☐	
[2]	[2]	March ☐		2009 ☐	
[3]	[3]	April ☐		2010 ☐	
	[4]	May ☐		2011 ☐	
	[5]	June ☐		2012 ☐	
	[6]	July ☐		2013 ☐	
	[7]	August ☐		2014 ☐	
	[8]	September ☐		2015 ☐	
	[9]	October ☐		2016 ☐	
		November ☐		2017 ☐	
		December ☐		2018 ☐	

PUPIL NUMBER

[0]	[0]	[0]	[0]	[0]	[0]
[1]	[1]	[1]	[1]	[1]	[1]
[2]	[2]	[2]	[2]	[2]	[2]
[3]	[3]	[3]	[3]	[3]	[3]
[4]	[4]	[4]	[4]	[4]	[4]
[5]	[5]	[5]	[5]	[5]	[5]
[6]	[6]	[6]	[6]	[6]	[6]
[7]	[7]	[7]	[7]	[7]	[7]
[8]	[8]	[8]	[8]	[8]	[8]
[9]	[9]	[9]	[9]	[9]	[9]

SCHOOL NUMBER

[0]	[0]	[0]	[0]	[0]	[0]	[0]
[1]	[1]	[1]	[1]	[1]	[1]	[1]
[2]	[2]	[2]	[2]	[2]	[2]	[2]
[3]	[3]	[3]	[3]	[3]	[3]	[3]
[4]	[4]	[4]	[4]	[4]	[4]	[4]
[5]	[5]	[5]	[5]	[5]	[5]	[5]
[6]	[6]	[6]	[6]	[6]	[6]	[6]
[7]	[7]	[7]	[7]	[7]	[7]	[7]
[8]	[8]	[8]	[8]	[8]	[8]	[8]
[9]	[9]	[9]	[9]	[9]	[9]	[9]

Please mark like this ⊢⊣

1
- £2.43 ☐
- £3.14 ☐
- £2.55 ☐
- £2.15 ☐
- £2.27 ☐

2
- 20:58 ☐
- 21:03 ☐
- 21:38 ☐
- 20:47 ☐
- 22:06 ☐

3
- 0.2003 kg ☐
- 2.003 kg ☐
- 2.03 kg ☐
- 2.033 kg ☐
- 2.3 kg ☐

4
- $\frac{1}{4}$ of 112 ☐
- $8\frac{1}{3}$ ☐
- $\frac{5}{8}$ of 48 ☐
- 31% of 105 ☐
- 20% of 145 ☐

5
- 150° ☐
- 230° ☐
- 120° ☐
- 270° ☐
- 240° ☐

6
- 31578 ☐
- 28980 ☐
- 27400 ☐
- 32434 ☐
- 284500 ☐

7
- 20 g ☐
- 27 g ☐
- 40 g ☐
- 35 g ☐
- 21 g ☐

8
- 43°C ☐
- 52°C ☐
- 21°C ☐
- 34°C ☐
- −6°C ☐

9
- 24 ☐
- 30 ☐
- 28 ☐
- 38 ☐
- 44 ☐

10
- £13.02 ☐
- £8.88 ☐
- £1.24 ☐
- £3.44 ☐
- £2.60 ☐

11
- 49 ☐
- 45 ☐
- 48 ☐
- 37 ☐
- 41 ☐

12
- 25% ☐
- 33% ☐
- 21% ☐
- 20% ☐
- 38% ☐

13
- 475 cm ☐
- 4.6 cm ☐
- 450 cm ☐
- 510 cm ☐
- 600 cm ☐

14
- 2.9962 ☐
- 2.62 ☐
- 2.0062 ☐
- 2.612 ☐
- 2.64 ☐

15
- 111 ☐
- 191 ☐
- 207 ☐
- 314 ☐
- 228 ☐

16
- A ☐
- B ☐
- C ☐
- D ☐
- E ☐

17
- £3.05 ☐
- £3.72 ☐
- £3.88 ☐
- £3.96 ☐
- £3.50 ☐

18
- £18.84 ☐
- £15.76 ☐
- £16.66 ☐
- £19.04 ☐
- £19.44 ☐

19
- 67° ☐
- 76° ☐
- 57° ☐
- 45° ☐
- 82° ☐

20
- 400 ☐
- 378.4 ☐
- 378.1 ☐
- 378.04 ☐
- 378.14 ☐

21
- 262 ☐
- 195 ☐
- 273 ☐
- 259 ☐
- 287 ☐

22
- 84 m ☐
- 46 m ☐
- 7200 cm ☐
- 55 m ☐
- 1005 cm ☐

23
- 17 ☐
- 15 ☐
- 19 ☐
- 18 ☐
- 14 ☐

24
- 342 km ☐
- 34.2 km ☐
- 34200 km ☐
- 0.342 km ☐
- 0.0342 km ☐

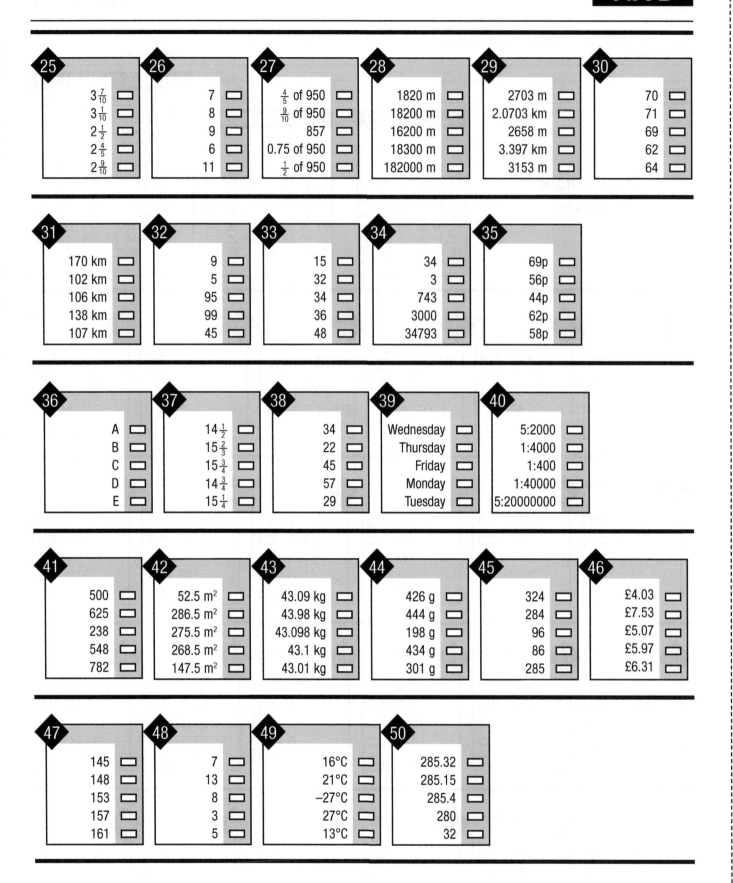

25
- $3\frac{7}{10}$ ☐
- $3\frac{1}{10}$ ☐
- $2\frac{1}{2}$ ☐
- $2\frac{4}{5}$ ☐
- $2\frac{9}{10}$ ☐

26
- 7 ☐
- 8 ☐
- 9 ☐
- 6 ☐
- 11 ☐

27
- $\frac{4}{5}$ of 950 ☐
- $\frac{9}{10}$ of 950 ☐
- 857 ☐
- 0.75 of 950 ☐
- $\frac{1}{2}$ of 950 ☐

28
- 1820 m ☐
- 18200 m ☐
- 16200 m ☐
- 18300 m ☐
- 182000 m ☐

29
- 2703 m ☐
- 2.0703 km ☐
- 2658 m ☐
- 3.397 km ☐
- 3153 m ☐

30
- 70 ☐
- 71 ☐
- 69 ☐
- 62 ☐
- 64 ☐

31
- 170 km ☐
- 102 km ☐
- 106 km ☐
- 138 km ☐
- 107 km ☐

32
- 9 ☐
- 5 ☐
- 95 ☐
- 99 ☐
- 45 ☐

33
- 15 ☐
- 32 ☐
- 34 ☐
- 36 ☐
- 48 ☐

34
- 34 ☐
- 3 ☐
- 743 ☐
- 3000 ☐
- 34793 ☐

35
- 69p ☐
- 56p ☐
- 44p ☐
- 62p ☐
- 58p ☐

36
- A ☐
- B ☐
- C ☐
- D ☐
- E ☐

37
- $14\frac{1}{2}$ ☐
- $15\frac{2}{3}$ ☐
- $15\frac{3}{4}$ ☐
- $14\frac{3}{4}$ ☐
- $15\frac{1}{4}$ ☐

38
- 34 ☐
- 22 ☐
- 45 ☐
- 57 ☐
- 29 ☐

39
- Wednesday ☐
- Thursday ☐
- Friday ☐
- Monday ☐
- Tuesday ☐

40
- 5:2000 ☐
- 1:4000 ☐
- 1:400 ☐
- 1:40000 ☐
- 5:20000000 ☐

41
- 500 ☐
- 625 ☐
- 238 ☐
- 548 ☐
- 782 ☐

42
- 52.5 m² ☐
- 286.5 m² ☐
- 275.5 m² ☐
- 268.5 m² ☐
- 147.5 m² ☐

43
- 43.09 kg ☐
- 43.98 kg ☐
- 43.098 kg ☐
- 43.1 kg ☐
- 43.01 kg ☐

44
- 426 g ☐
- 444 g ☐
- 198 g ☐
- 434 g ☐
- 301 g ☐

45
- 324 ☐
- 284 ☐
- 96 ☐
- 86 ☐
- 285 ☐

46
- £4.03 ☐
- £7.53 ☐
- £5.07 ☐
- £5.97 ☐
- £6.31 ☐

47
- 145 ☐
- 148 ☐
- 153 ☐
- 157 ☐
- 161 ☐

48
- 7 ☐
- 13 ☐
- 8 ☐
- 3 ☐
- 5 ☐

49
- 16°C ☐
- 21°C ☐
- −27°C ☐
- 27°C ☐
- 13°C ☐

50
- 285.32 ☐
- 285.15 ☐
- 285.4 ☐
- 280 ☐
- 32 ☐

MA C

Pupil's Name

School Name

Date of Test

		DATE OF BIRTH	
	Day	Month	Year
	[0] [0]	January ☐	2007 ☐
	[1] [1]	February ☐	2008 ☐
	[2] [2]	March ☐	2009 ☐
	[3] [3]	April ☐	2010 ☐
	[4]	May ☐	2011 ☐
	[5]	June ☐	2012 ☐
	[6]	July ☐	2013 ☐
	[7]	August ☐	2014 ☐
	[8]	September ☐	2015 ☐
	[9]	October ☐	2016 ☐
		November ☐	2017 ☐
		December ☐	2018 ☐

PUPIL NUMBER

[0]	[0]	[0]	[0]	[0]	[0]
[1]	[1]	[1]	[1]	[1]	[1]
[2]	[2]	[2]	[2]	[2]	[2]
[3]	[3]	[3]	[3]	[3]	[3]
[4]	[4]	[4]	[4]	[4]	[4]
[5]	[5]	[5]	[5]	[5]	[5]
[6]	[6]	[6]	[6]	[6]	[6]
[7]	[7]	[7]	[7]	[7]	[7]
[8]	[8]	[8]	[8]	[8]	[8]
[9]	[9]	[9]	[9]	[9]	[9]

SCHOOL NUMBER

[0]	[0]	[0]	[0]	[0]	[0]	[0]
[1]	[1]	[1]	[1]	[1]	[1]	[1]
[2]	[2]	[2]	[2]	[2]	[2]	[2]
[3]	[3]	[3]	[3]	[3]	[3]	[3]
[4]	[4]	[4]	[4]	[4]	[4]	[4]
[5]	[5]	[5]	[5]	[5]	[5]	[5]
[6]	[6]	[6]	[6]	[6]	[6]	[6]
[7]	[7]	[7]	[7]	[7]	[7]	[7]
[8]	[8]	[8]	[8]	[8]	[8]	[8]
[9]	[9]	[9]	[9]	[9]	[9]	[9]

Please mark like this ⊢

1
- 103400 ☐
- 1050000 ☐
- 175000 ☐
- 1750000 ☐
- 1705000 ☐

2
- 16 ☐
- 32 ☐
- 26 ☐
- 40 ☐
- 27 ☐

3
- £28.85 ☐
- £28.81 ☐
- £20.18 ☐
- £20.81 ☐
- £17.53 ☐

4
- 12° ☐
- −12° ☐
- −7° ☐
- 3° ☐
- −18° ☐

5
- 50300 mm ☐
- 604 mm ☐
- 283 mm ☐
- 951.4 mm ☐
- 855 mm ☐

6
- 0.63 ☐
- 0.063 ☐
- 63.1 ☐
- 6.3 ☐
- 0.363 ☐

7
- 4770 ☐
- 5765 ☐
- 5795 ☐
- 4281 ☐
- 3666 ☐

8
- 76% ☐
- 77% ☐
- 68% ☐
- 83% ☐
- 85% ☐

9
- 175 ☐
- 210 ☐
- 435 ☐
- 245 ☐
- 3500 ☐

10
- 78 ☐
- 84 ☐
- 91 ☐
- 92 ☐
- 85 ☐

11
- 97 ☐
- 81 ☐
- 75 ☐
- 102 ☐
- 84 ☐

12
- 115 ☐
- 121 ☐
- 181 ☐
- 126 ☐
- 144 ☐

13
- 14.19 sec ☐
- 13.794 sec ☐
- 13.57 sec ☐
- 14.37 sec ☐
- 13.47 sec ☐

14
- 192 m² ☐
- 178 m² ☐
- 168 m² ☐
- 204 m² ☐
- 185 m² ☐

15
- 542 ☐
- 649 ☐
- 487 ☐
- 546 ☐
- 586 ☐

16
- 4216 ☐
- 3968 ☐
- 3986 ☐
- 2574 ☐
- 5012 ☐

17
- £7.56 ☐
- £30.24 ☐
- £34.44 ☐
- £39.76 ☐
- £16.54 ☐

18
- 3 ☐
- 4 ☐
- 2 ☐
- 1 ☐
- 5 ☐

19
- 13129 ☐
- 15256 ☐
- 13149 ☐
- 18302 ☐
- 25601 ☐

20
- 1246.88 m³ ☐
- 128.664 m³ ☐
- 12847.45 m³ ☐
- 128.466 m³ ☐
- 12846600 m³ ☐

21
- A ☐
- B ☐
- C ☐
- D ☐
- E ☐

22
- $\frac{5}{8}$ ☐
- $\frac{6}{9}$ ☐
- $\frac{4}{7}$ ☐
- 0.6 ☐
- $\frac{12}{15}$ ☐

23
- 568 ml ☐
- 634 ml ☐
- 589 ml ☐
- 632 ml ☐
- 726 ml ☐

24
- 79.1 kg ☐
- 78.899 kg ☐
- 79.087 kg ☐
- 78.9027 kg ☐
- 78.909 kg ☐

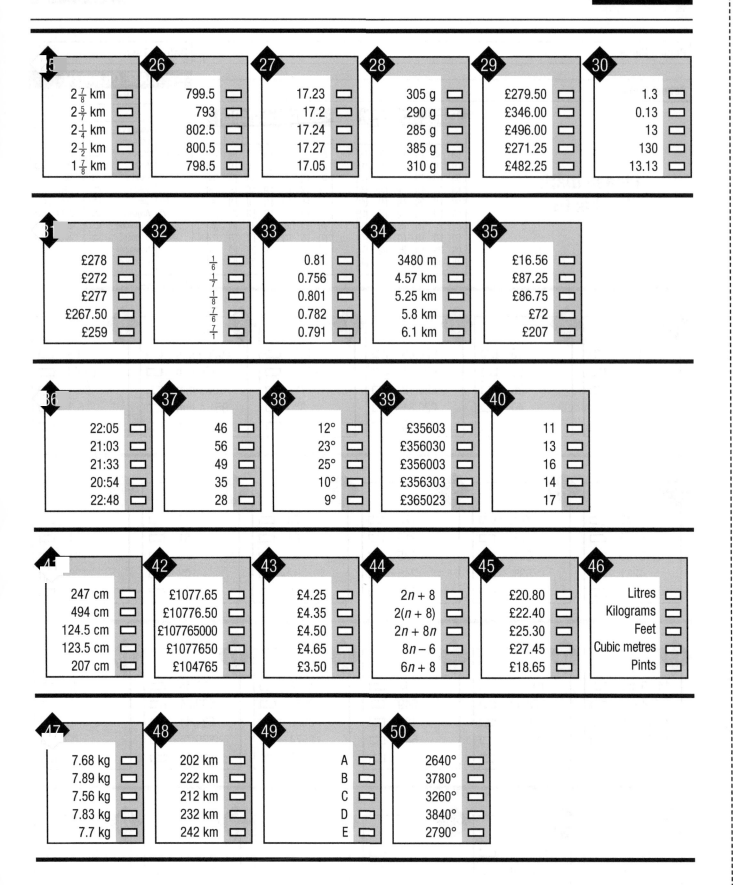

25
- $2\frac{7}{8}$ km ▢
- $2\frac{5}{7}$ km ▢
- $2\frac{1}{4}$ km ▢
- $2\frac{1}{2}$ km ▢
- $1\frac{7}{8}$ km ▢

26
- 799.5 ▢
- 793 ▢
- 802.5 ▢
- 800.5 ▢
- 798.5 ▢

27
- 17.23 ▢
- 17.2 ▢
- 17.24 ▢
- 17.27 ▢
- 17.05 ▢

28
- 305 g ▢
- 290 g ▢
- 285 g ▢
- 385 g ▢
- 310 g ▢

29
- £279.50 ▢
- £346.00 ▢
- £496.00 ▢
- £271.25 ▢
- £482.25 ▢

30
- 1.3 ▢
- 0.13 ▢
- 13 ▢
- 130 ▢
- 13.13 ▢

31
- £278 ▢
- £272 ▢
- £277 ▢
- £267.50 ▢
- £259 ▢

32
- $\frac{1}{6}$ ▢
- $\frac{1}{7}$ ▢
- $\frac{1}{8}$ ▢
- $\frac{7}{6}$ ▢
- $\frac{7}{1}$ ▢

33
- 0.81 ▢
- 0.756 ▢
- 0.801 ▢
- 0.782 ▢
- 0.791 ▢

34
- 3480 m ▢
- 4.57 km ▢
- 5.25 km ▢
- 5.8 km ▢
- 6.1 km ▢

35
- £16.56 ▢
- £87.25 ▢
- £86.75 ▢
- £72 ▢
- £207 ▢

36
- 22:05 ▢
- 21:03 ▢
- 21:33 ▢
- 20:54 ▢
- 22:48 ▢

37
- 46 ▢
- 56 ▢
- 49 ▢
- 35 ▢
- 28 ▢

38
- 12° ▢
- 23° ▢
- 25° ▢
- 10° ▢
- 9° ▢

39
- £35603 ▢
- £356030 ▢
- £356003 ▢
- £356303 ▢
- £365023 ▢

40
- 11 ▢
- 13 ▢
- 16 ▢
- 14 ▢
- 17 ▢

41
- 247 cm ▢
- 494 cm ▢
- 124.5 cm ▢
- 123.5 cm ▢
- 207 cm ▢

42
- £1077.65 ▢
- £10776.50 ▢
- £107765000 ▢
- £1077650 ▢
- £104765 ▢

43
- £4.25 ▢
- £4.35 ▢
- £4.50 ▢
- £4.65 ▢
- £3.50 ▢

44
- $2n + 8$ ▢
- $2(n + 8)$ ▢
- $2n + 8n$ ▢
- $8n - 6$ ▢
- $6n + 8$ ▢

45
- £20.80 ▢
- £22.40 ▢
- £25.30 ▢
- £27.45 ▢
- £18.65 ▢

46
- Litres ▢
- Kilograms ▢
- Feet ▢
- Cubic metres ▢
- Pints ▢

47
- 7.68 kg ▢
- 7.89 kg ▢
- 7.56 kg ▢
- 7.83 kg ▢
- 7.7 kg ▢

48
- 202 km ▢
- 222 km ▢
- 212 km ▢
- 232 km ▢
- 242 km ▢

49
- A ▢
- B ▢
- C ▢
- D ▢
- E ▢

50
- 2640° ▢
- 3780° ▢
- 3260° ▢
- 3840° ▢
- 2790° ▢

MA

Pupil's Name		Date of Test

School Name

DATE OF BIRTH

Day	Month	Year
[0] [0]	January	2007
[1] [1]	February	2008
[2] [2]	March	2009
[3] [3]	April	2010
	May	2011
[4]	June	2012
[5]	July	2013
[6]	August	2014
[7]	September	2015
[8]	October	2016
[9]	November	2017
	December	2018

PUPIL NUMBER

[0] [0] [0] [0] [0] [0]
[1] [1] [1] [1] [1] [1]
[2] [2] [2] [2] [2] [2]
[3] [3] [3] [3] [3] [3]
[4] [4] [4] [4] [4] [4]
[5] [5] [5] [5] [5] [5]
[6] [6] [6] [6] [6] [6]
[7] [7] [7] [7] [7] [7]
[8] [8] [8] [8] [8] [8]
[9] [9] [9] [9] [9] [9]

SCHOOL NUMBER

[0] [0] [0] [0] [0] [0] [0]
[1] [1] [1] [1] [1] [1] [1]
[2] [2] [2] [2] [2] [2] [2]
[3] [3] [3] [3] [3] [3] [3]
[4] [4] [4] [4] [4] [4] [4]
[5] [5] [5] [5] [5] [5] [5]
[6] [6] [6] [6] [6] [6] [6]
[7] [7] [7] [7] [7] [7] [7]
[8] [8] [8] [8] [8] [8] [8]
[9] [9] [9] [9] [9] [9] [9]

Please mark like this ⊢

1
- 8624
- 86240
- 860024
- 86024
- 80624

2
- 25
- 22
- 34
- 26
- 31

3
- 9:25
- 21:25
- 19:35
- 22:35
- 20:35

4
- 427
- 127
- 428
- 113
- 159

5
- 28.8
- 28.08
- 2880
- 28080
- 280800

6
- 0.45 g
- 4.5 g
- 3.05 g
- 1.005 g
- 5.4 g

7
- 1350 g
- 1250 g
- 675 g
- 900 g
- 1125 g

8
- 0.1
- 0.5
- 0.15
- 0.05
- 0.005

9
- 25 g
- 28 g
- 35 g
- 40 g
- 55 g

10
- 0.974
- 74
- 974
- 84
- 884

11
- Friday
- Sunday
- Wednesday
- Monday
- Saturday

12
- 306
- 32
- 144
- 59
- 45

13
- 11
- 4
- 7
- 9
- 8

14
- 17 min
- 23 min
- 36 min
- 28 min
- 31 min

15
- $x = 2y$
- $x = y - 2$
- $x = 13$
- $x = 2$
- $x = y + 2$

16
- 1:40
- 1:4
- 1:400
- 1:4000
- 1:7

17
- Monday
- Tuesday
- Wednesday
- Thursday
- Friday

18
- 40 km
- 44 km
- 50 km
- 66 km
- 63 km

19
- 24
- 23
- 17
- 11
- 15

20
- £145.37
- £143.47
- £146.27
- £145.27
- £144.37

21
- $4b + 6b$
- $12b$
- $6b + 8c$
- $14b$
- $10b$

22
- 0.00001
- 100
- 0.001
- 1000
- 0.0001

23
- 8
- 7
- 15
- 80
- 56

24
- £29.86
- £28.21
- £25.12
- £20.14
- £22.14

...bers Ltd

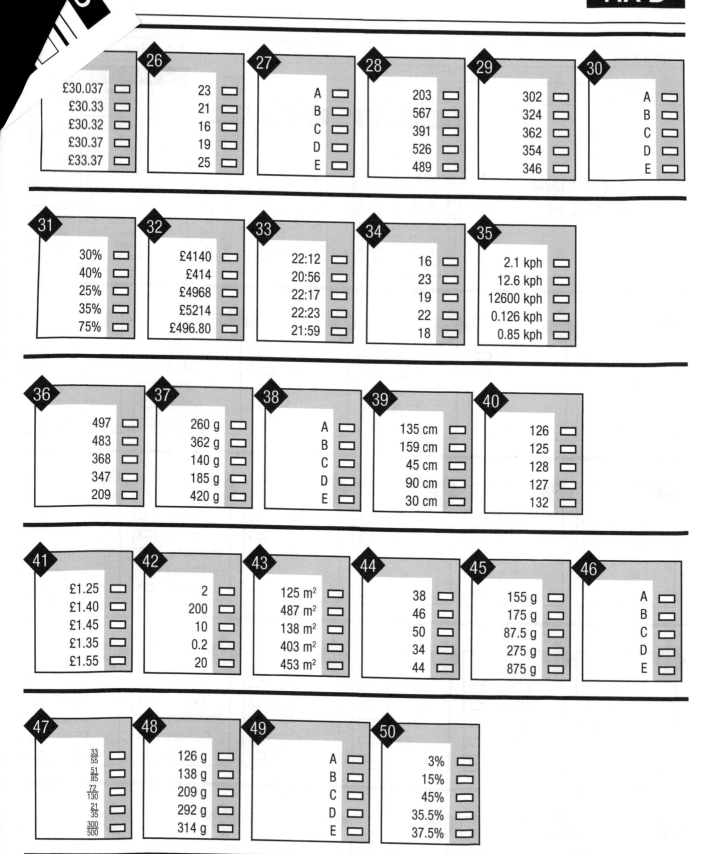

26
- £30.037 ☐
- £30.33 ☐
- £30.32 ☐
- £30.37 ☐
- £33.37 ☐

- 23 ☐
- 21 ☐
- 16 ☐
- 19 ☐
- 25 ☐

27
- A ☐
- B ☐
- C ☐
- D ☐
- E ☐

28
- 203 ☐
- 567 ☐
- 391 ☐
- 526 ☐
- 489 ☐

29
- 302 ☐
- 324 ☐
- 362 ☐
- 354 ☐
- 346 ☐

30
- A ☐
- B ☐
- C ☐
- D ☐
- E ☐

31
- 30% ☐
- 40% ☐
- 25% ☐
- 35% ☐
- 75% ☐

32
- £4140 ☐
- £414 ☐
- £4968 ☐
- £5214 ☐
- £496.80 ☐

33
- 22:12 ☐
- 20:56 ☐
- 22:17 ☐
- 22:23 ☐
- 21:59 ☐

34
- 16 ☐
- 23 ☐
- 19 ☐
- 22 ☐
- 18 ☐

35
- 2.1 kph ☐
- 12.6 kph ☐
- 12600 kph ☐
- 0.126 kph ☐
- 0.85 kph ☐

36
- 497 ☐
- 483 ☐
- 368 ☐
- 347 ☐
- 209 ☐

37
- 260 g ☐
- 362 g ☐
- 140 g ☐
- 185 g ☐
- 420 g ☐

38
- A ☐
- B ☐
- C ☐
- D ☐
- E ☐

39
- 135 cm ☐
- 159 cm ☐
- 45 cm ☐
- 90 cm ☐
- 30 cm ☐

40
- 126 ☐
- 125 ☐
- 128 ☐
- 127 ☐
- 132 ☐

41
- £1.25 ☐
- £1.40 ☐
- £1.45 ☐
- £1.35 ☐
- £1.55 ☐

42
- 2 ☐
- 200 ☐
- 10 ☐
- 0.2 ☐
- 20 ☐

43
- 125 m² ☐
- 487 m² ☐
- 138 m² ☐
- 403 m² ☐
- 453 m² ☐

44
- 38 ☐
- 46 ☐
- 50 ☐
- 34 ☐
- 44 ☐

45
- 155 g ☐
- 175 g ☐
- 87.5 g ☐
- 275 g ☐
- 875 g ☐

46
- A ☐
- B ☐
- C ☐
- D ☐
- E ☐

47
- $\frac{33}{55}$ ☐
- $\frac{51}{85}$ ☐
- $\frac{72}{130}$ ☐
- $\frac{21}{35}$ ☐
- $\frac{300}{500}$ ☐

48
- 126 g ☐
- 138 g ☐
- 209 g ☐
- 292 g ☐
- 314 g ☐

49
- A ☐
- B ☐
- C ☐
- D ☐
- E ☐

50
- 3% ☐
- 15% ☐
- 45% ☐
- 35.5% ☐
- 37.5% ☐